DISCOVERING RETROVIRUSES

DISCOVERING RETROVIRUSES

Beacons in the Biosphere

ANNA MARIE SKALKA

Harvard University Press

Cambridge, Massachusetts

London, England

2018

Library of Congress Cataloging-in-Publication Data

Names: Skalka, Anna Marie, author.
Title: Discovering retroviruses : beacons in the biosphere / Anna Marie Skalka.
Description: Cambridge, Massachusetts : Harvard University Press, 2018. |
    Includes bibliographical references and index.
Identifiers: LCCN 2018002504 | ISBN 9780674971707 (hardcover : alk. paper)
Subjects: LCSH: Retroviruses. | Retrovirus infections. | Viruses—Evolution. |
    Medicine—Research—History.
Classification: LCC QR414.5 S53 2018 | DDC 579.2/569—dc23
LC record available at https://lccn.loc.gov/2018002504

This book was written for readers of all ages who marvel at the wonders of our biologic cosmos and at the manner in which they have been revealed.

It is dedicated to my husband, Rudy; children, Jeannemarie and Christian; and grandchildren, Kazimir and Shiloh, that they may glimpse the beacons that have lit my way.

— *Contents* —

— *Tables and Figures* —

DISCOVERING RETROVIRUSES

— *Introduction* —

Retroviruses are ancient invaders of their host's genomes. This virus family is unique among animal viruses in its capacity to copy its genetic blueprint of RNA into DNA (the hereditary material in all animals) and then to splice that copy into the DNA of its host cell. While the host cell normally survives this invasion, the retroviral DNA remains as a permanent genetic parasite, co-opting cellular machinery to maintain its existence and, in most cases, to produce new retrovirus particles that can infect other cells. About 8 percent of the DNA in our bodies comprises copies of ancient retroviral DNA and bits thereof, whereas only 1.5 percent accounts for information that encodes *all* of the proteins that make up our cells and ensure their function. This dichotomy came as quite a shock to the scientists who sequenced the first human genome—a somewhat humbling realization that retroviruses "R" us.

But we are not alone; retroviral DNA or that of its predecessors is present in all branches of the tree of life. In the course of evolution, multicellular organisms have acquired numerous mechanisms to defend against retroviral invasion—in an ongoing host versus virus "arms race." Nevertheless, in some notable cases, acquisition of retroviral genes has actually afforded an opportunity for its host. Reprogramming of retroviral genes in host DNA launched the evolution of placental mammals, to which we humans belong.

Retroviruses are often, but not always, benign modern travelers. Retroviruses can cause a variety of cancers in humans and other animals. Analysis of the mechanisms by which they do so has revealed the genetic basis of this disease. On the other hand, the retroviral, HIV-induced AIDS has killed more people worldwide than any other disease of viral origin. While there have been tremendous strides in the delineation of protective measures, and ever-improving treatments are available, HIV / AIDS is still an incurable disease. It may therefore be surprising to learn that genetically gutted versions of HIV and its relatives are being employed as vehicles for gene therapy and to better understand and treat AIDS and other diseases. Human beings have found ways to advance both science and medicine by harnessing even this horrific killer.

The pages that follow describe the amazing stories of the discovery of these ancient invaders / modern travelers and the fascinating "yin and yang" of their existence. They will also disclose how studies of retroviruses and concomitant discoveries in genetics have illuminated the landscape of our biological world. Traveling back in time to the origin of life on Earth, through evolution and to the present day, we will track the many astonishing revelations garnered through pursuit of these unique *Beacons in the Biosphere*.

Early Pioneers

By the end of the nineteenth century, tiny organisms invisible to the naked eye, such as bacteria, had already been seen and studied with light microscopes. Furthermore, a connection between bacteria and disease had been established by the German physician Robert Koch in his studies of anthrax, a common disease of cattle at the time. Viruses, however, were mysterious entities and an entirely new kind of disease agent. The first of these to be described, in 1892, was the plant virus that causes tobacco mosaic disease. The second, discovered four years later, induces foot-and-mouth disease in cattle. Unlike bacteria, these plant and animal pathogens could *not* be seen in the existing microscopes and, indeed, were so small that they could pass through filters known to retain bacteria and other microbes. Most important, these agents couldn't be propagated in broth solutions like bacteria; their numbers increased only within the cells of their host organism. Because these new "filterable" agents seemed like some sort of contagious liquid poison to early researchers, they were given the Latin name for poison: viruses.

The very first retrovirus discovered is actually related to our modern scourge, the AIDS virus, HIV (human immunodeficiency virus). While studying an anemia in horses called swamp fever in the early 1900s, two French veterinarians, Henri Vallée and Henri Carré, found that a filterable agent now known as the equine infectious anemia virus (EIAV) could pass the disease from one animal to another. Four years later, two Danish scientists, Vilhelm Ellerman and Olaf Bang, described another filterable agent, one that could transmit leukemia among chickens. But these two early accounts of what would eventually be identified as retroviral infections made little impression on the medical community. A disease of horses seemed irrelevant to humans, and leukemias were not recognized as cancer at the time. This all changed when an American

Table 1.1 Selected Timelines for Discoveries in Genetics and Retrovirology

1866 Gregor Mendel reports results of studies of inheritance in pea plants.

1869 Friedrich Miescher identifies DNA (first called "nuclein") in cell nuclei.

1882 Walther Flemming reports visualization of chromosome doubling and movement during cell division.

1900 Mendel's experiments are "rediscovered" by three botanists: a Dutchman, Hugo de Vries; a German, Carl Correns; and an Austrian, Erich Tschermak.

1902–1903 Theodor Boveri and Walter Sutton propose the chromosome theory of heredity, which conforms to Mendelian principles.

1904 French veterinarians Henri Vallée and Henri Carré discover a filterable agent that transmits anemia in horses, now classified as a retrovirus, equine infectious anemia virus (EIAV).

1905 The British scientist, William Bateson, translates Mendel's paper into English and names the field of study "genetics."

1908 Danish scientists Vilhelm Ellerman and Olaf Bang discover an infectious filterable agent that causes leukemia in chickens, later identified as the retrovirus avian leukemia virus (ALV).

1910–1911 T. H. Morgan and colleagues confirm the chromosome theory of heredity in studies of sex-linked inheritance of eye color in fruit flies.

1911 Peyton Rous discovers a tumor-inducing infectious agent in chickens. The agent was later known as the Rous sarcoma virus and assigned to the avian sarcoma virus (ASV) group of retroviruses.

1913 Alfred Sturtevant (a student working with T. H. Morgan) constructs the first gene linkage map, showing that genes are arranged like beads on a string in chromosomes.

1914 Rous, as well as Fujinami and Inamoto, independently report additional virus-induced tumors. The Japanese isolate is later known as Fujinami sarcoma virus (FSV).

1914–1918 World War I

1928 British bacteriologist Fred Griffith reports that some component from heat-killed virulent pneumococcal bacteria can "transform" a nonvirulent strain into a virulent one.

1931 Harriet Creighton and Barbara McClintock show that genetic recombination is associated with the physical exchange of chromosomal segments.

1936 John Joseph Bittner discovers a "milk factor" that is transmitted by cancerous mothers to young mice while nursing, later identified as the retrovirus mouse mammary tumor virus (MMTV).

1939–1945 World War II

1944 Oswald Avery, William MacLeod, and Maclyn McCarty report that DNA from virulent strains mediates transformation of nonvirulent strains of *Pneumococcus.*

1950 Erwin Chargaff reports that DNA contains equal amounts of the bases A and T and of C and G, although the ratios of A+T to C+G ratio can differ between organisms.

1951–1957 Ludwik Gross, at the veterans hospital in New York City, isolates a potent virus that causes mouse leukemia, mouse leukemia virus (MLV), later assigned to the retrovirus family.

1952 Alfred Hershey and Martha Chase confirm that DNA mediates heredity in bacteriophage labeling experiments.

1953 James Watson and Francis Crick report their model of the DNA double-helix structure.

Table 1.1 (continued)

1954 George Palade observed ribosomes in the cytoplasm of cells by using an electron microscope. These dense particles were later shown to be protein synthesis factories.
1954 Bjorn Sigurdsson describes an infectious agent that causes a slowly developing but fatal disease in sheep in Iceland. Called Visna virus, it is later shown to be related to the AIDS retrovirus, HIV.
1956 Arthur Kornberg and colleagues isolate an enzyme from the bacterium, *Escherichia coli*, that copies a template DNA strand.
1957 Charlotte Friend reports that cell-free transmission of leukemia in mice is caused by a virus to be known as Friend murine leukemia virus (Fr-MLV). This was one of many new retroviruses to be discovered in the following years.
1958 Matthew Meselson and Franklin Stahl DNA show that DNA replication is semi-conservative, using a technique they invented called "equilibrium density gradient centrifugation," which relied on the use of "heavy" isotopes.
1958 Paul C. Zamecnik and Mahlon Hoagland discover "adapter" RNAs with bound single amino acids, later called transfer RNAs (tRNAs).
1960 Sidney Brenner, Francis Crick, François Jacob, and Jacques Monod predict that messenger RNA (mRNA) is the intermediate between DNA and protein.
1961 The genetic code is cracked by Marshall Nirenberg, together with colleagues Matthaei and Leder, and by Har Gobind Khorana.

Note: Bold dates identify discoveries that were recognized by Nobel Prizes.

pathologist, Francis Peyton Rous, made the startling announcement that a filterable agent could cause solid tumors.

Peyton Rous was trained as a physician at Johns Hopkins Medical School in Baltimore but decided after his internship that taking care of patients was neither his forte nor calling. Feeling "unfit" for clinical medicine, he obtained an assistantship in pathology at the University of Michigan, where he could focus instead on laboratory research. The story of his discovery of a virus that could cause solid tumors began in 1909 when, only four years after receiving his medical degree, he was invited to set up a cancer research laboratory at the Rockefeller Institute for Medical Research in New York (now Rockefeller University). Shortly after establishing his new laboratory, Rous was visited by a woman from a nearby poultry farm who brought a hen with a large lump in its breast for examination. Rous recognized the lump as a sarcoma (a tumor of connective tissue), and perhaps because he was new to the profession with no baggage of biases, this hen sparked his interest. Rous was aware of earlier reports that cancers can sometimes spread among animals, much like an infection, and he decided to test this idea. He first observed that new sarcomas were formed when he placed bits of the tumor in the bodies of healthy chickens from the same inbred stock. To find out if such spread was due to a virus, he mixed ground bits of the tumor in a saline solution and then passed the solution through a filter that would exclude bacteria. When new tumors arose in chickens injected with the filtered solution, Rous concluded that the infectious agent must certainly be a virus and reported his findings in two papers in 1910 and 1911.[1]

As is often the case with results that are unexpected or disrupt current paradigms, Rous's work generated considerable controversy. Some of his contemporaries argued that the filtrates were probably not cell free or that the sarcomas were not really cancer. Others considered chickens to be too distantly related to provide a useful model for human cancer. Indeed, the very idea that one could "catch" cancer from such a tiny agent was met with general skepticism. Discouraged by the response of his peers, Rous abandoned this line of research for decades, focusing instead on World War I–related projects. His wartime work was also quite productive; Rous developed the blood-preserving techniques that led to establishment of the world's first blood bank near the front line in Belgium. After the war, he returned to study cancer caused by another virus, the Shope papilloma virus of rabbits. But, as we will see in later chapters, discovery of the retrovirus named after him, the Rous sarcoma virus (RSV), was to be considered his crowning achievement, as it laid the foundation for a broad understanding of both retroviruses and cancer. In 1966, Rous was awarded the Nobel Prize for this work—three years before his death at the age of ninety.

Although reports of similar cancer-causing agents of mice and other animal species occurred sporadically in the years following Rous's initial discovery, it would be decades until the unique properties of retroviruses were appreciated. Such recognition was to await the development of experimental tools and insights from research with viruses that infect bacteria, called bacteriophages (or simply phages; from the Greek word to eat, "phagos"), and the establishment of the new fields of genetics and molecular biology.

To fully understand the impact of retroviruses on the fields of genetics, evolution, and medicine, we must look at the history of research into the molecular basis of heredity, a story unfolding at the very same time that the first retroviruses were being discovered.

Genes, DNA, and Establishment of the "Central Dogma"

In today's world, almost everybody knows about genes and DNA, if not from news reports, then certainly from popular television shows, where a murderer's identity can be determined from a small, dried drop of blood found at the crime scene. For this reason, it may be hard to imagine that it was not known until the middle of the past century that heritable traits (genes) are encoded in DNA. The fascinating path to this knowledge was somewhat circuitous and progress extremely slow. Indeed, the first critical observations were made in the middle of the 1800s by Gregor Mendel, an Austrian monk and scientist, working in the experimental garden of an Augustinian abbey in what is now the Czech Republic.

Mendel chose pea plants as an experimental system to study heredity because many distinct varieties were available, and offspring could be produced quickly and easily in the abbey garden. He was interested in knowing how particular traits, such as seed color or shape, were passed on through generations. Over a period of eight years, while performing meticulous crosses and studying almost 30,000 plants, Mendel was able to establish several fundamental principles. First, he demonstrated that colors and shapes were inherited independently of one another, as if they were specified by separate "units." He also determined that when purebred plants were crossed, the property of one parent (e.g., a particular color) could be dominant in the next generation of hybrid plants. Results from his numerous crosses between these hybrid plants showed Mendel that the property that seemed to disappear in the hybrids, called the "recessive" trait, would reappear among their progeny—and, importantly, in *particular* ratios. On average, one plant exhibited the recessive property for every three with the dominant property (1:3). By

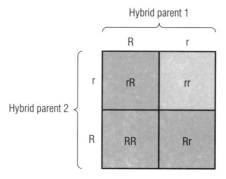

Fig. 1.1 Mendel's crosses depicted in a Punnett square. The Punnett square, named after the geneticist Reginald Crundall Punnett (1875–1967), is used by biologists to determine the probability that an offspring will exhibit a particular trait as shown in the boxes. Mendel deduced that genetic traits must exist in pairs, one from each parent. When one of a pair is dominant over the other (e.g., green over yellow), inheritance from a cross of hybrid parents can be described by the ratio of 1:2:1 of pure recessive (rr), hybrid (rR or Rr), and pure dominant (RR). As R is dominant over r, the ratio of green to yellow progeny produced by the hybrid parents is 3:1.

applying statistical methods and mathematical models, most unusual for a botanist at the time, Mendel deduced that the relationship between the dominant and recessive properties could be explained if two independent trait-defining units existed in each parent, either one of which was passed on to the next generation. Inheritance could be described by the ratio of 1:2:1 of recessive, hybrid, and dominant, as illustrated in the boxes of a Punnett square. This is a phenomenon with which we are now familiar in human eye color; that is, two parents with brown eyes (the dominant trait) can produce a child with blue eyes if they both carry and pass on their recessive gene for blue eyes.

A somewhat modest and deferential scientist, Mendel was discouraged when a leading authority of the time, the Swiss botanist Karl Willhelm von Nägeli, rejected Mendel's results and conclusions. Nägeli had his own ideas for how genetic traits are inherited (now known to be incorrect), and he was writing a book on this topic. He didn't believe Mendel's results to be generally applicable because, as Mendel later confirmed, the plant Nägeli was studying did not transmit genetic traits in the same way as Mendel's peas. What neither scientist knew at the time was that this difference was due to the fact that Nägeli's plants reproduce mainly via a mechanism in which the seeds carry genetic traits of only one parent. Mendel eventually delivered two lectures on his findings at the 1865 meeting of the Natural Science Society in Brno, and his results were published in its journal the following year. Nevertheless, although he is now universally recognized as the founder of genetics, Mendel's work was overlooked for the next forty years. The principles established by his experiments, now known as Mendel's laws of inheritance ("independent assortment," "segregation," and "dominance"), were rediscovered in 1900 (sixteen years after Mendel died) when other researchers, using peas and other plants, sought to explain their similar findings. As happens not infrequently in science, discoveries that seem odd or even irrelevant at the time they are made can provide essential pieces to important puzzles that arise decades later. This posthumous appreciation of Mendel's work inspired a greatly renewed interest in genetics, but it would take another fifty years until we would know what genes are made of.

By the late 1800s, use of the light microscope allowed the visualization not only of some microorganisms but also of features inside of cells. Among these were linear structures within the nuclei, called chromosomes (from the Greek words for colored, "chroma," and bodies, "soma"), which could be seen most clearly when cells were in the process of dividing. The realization that genes reside within these linear chromosomes came from independent studies in 1902–1903 by Walter Sutton, a graduate student at Columbia University, and an established German scientist, Theodor Boveri. Sutton based his theory on observations of insect chromosomes, which are lined up in matched pairs that separate during formation of eggs and sperm, precisely as expected by Mendel's laws. Boveri's studies showed that a proper complement of chromosomes had to be present if the sea urchin embryos he was studying were to develop normally. The chromosome theory of inheritance gained further support in studies of fruit fly eye color by Thomas Hunt Morgan and was accepted broadly in 1931 when Harriet Creighton and Barbara McClintock showed that genetic recombination is correlated with physical exchanges between chromosomes in dividing corn cells.

Chromosomes were known to contain both proteins and DNA, but it was the protein component that was assumed to comprise genes in the 1930s. Consisting of long chains with different combinations of twenty amino acids, proteins appeared to be the only cellular constituents sufficiently complex to embody hereditary traits. DNA, made up of long, repetitious chains of only four different nucleotides (sugar-phosphate molecules, which contain one of four bases), was considered a most unlikely candidate. Consequently, the best guess at the time was that DNA was some sort of general "scaffold" that supported genes.

The first evidence that DNA is heredity material was obtained in 1944 by Oswald Avery and his Rockefeller University colleagues, Colin MacLeod and Maclyn McCarty. Avery became intrigued by a 1928 report that a component from killed pathogenic pneumococcal bacteria could induce virulence in a benign strain of the same bacteria when the two were mixed together, a process called "transformation." He and his colleagues decided to track down the responsible component by a process of elimination. They prepared extracts from the killed virulent strain and treated the extracts with enzymes that would destroy proteins, lipids, or DNA. They then exposed the benign bacteria to the extracts to discover which treatments abolished transformation, which they monitored by the formation of large, smooth colonies of virulent bacteria. Their experiments showed that transformation activity was lost only after treatment with enzymes able to destroy DNA. Furthermore, addition of chemically purified DNA from the virulent strain caused transformation of cultured benign cells. From these results, Avery and his colleagues concluded that genes specifying virulence must be encoded in DNA. As one may suspect, their seemingly compelling results were met not with acclaim but with general disbelief. Critics grumbled that Avery's DNA preparations might be contaminated with sufficient amounts of protein to cause the resulting transformation. A Rockefeller colleague, Alfred Mirsky, copublished a widely read, critical article in 1946 noting that "as much as 1 or 2 percent of protein could be present in a preparation of 'pure, protein-free' nucleic acid."[2] Indeed, the idea that genes were embodied in proteins was so thoroughly accepted at the time, many leading biologists simply disregarded Avery's work.

The situation changed in 1952 with a publication from Alfred Hershey's laboratory at the Carnegie Institution's Department of Genetics

10

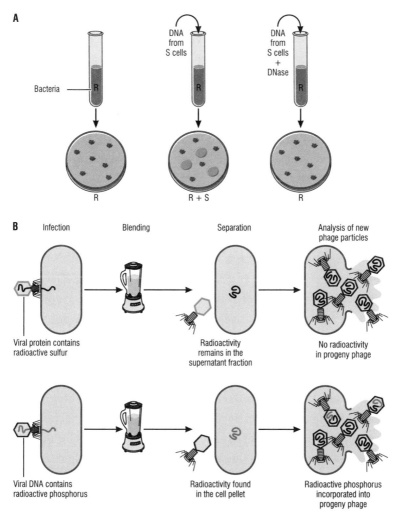

Fig. 1.2 Genes are encoded in DNA. (A) An illustration of the 1944 experiment of Avery, McLeod, and McCarty. The nonvirulent pneumococcal bacteria suspended in the first test tube make small rough colonies when planted in culture dishes (R). A mixture of small rough and large smooth (S) colonies is formed if DNA purified from the virulent strain is added to the nonvirulent bacteria, shown in the center. As illustrated on the right, only rough colonies are formed if the purified DNA from the virulent strain is first treated with an enzyme (DNase) known to degrade DNA. **(B)** An illustration of the Hershey-Chase blender experiment. Radioactive protein in an infecting phage particle is shown as an orange head, and radioactive DNA injected from an infecting particle is red. Only the DNA enters infected cells and is incorporated into progeny phage particles.

in Cold Spring Harbor. Hershey was a founding member of a tight-knit cohort of scientists called the Phage Group. It had been known since 1915 that bacterial cells are also hosts to viruses (bacteriophages). Members of the Phage Group were convinced that bacteriophages offered the best chance at understanding the molecular basis of heredity because they likely had small, manageable sets of genes. The host bacterium of choice for the Phage Group was *Escherichia coli* (*E. coli*), a common constituent of the human gut and easy to culture in the laboratory. *E. coli* was to become a workhorse for the field and is today one of the best-understood organisms in the world. The *E. coli* phage that Hershey studied, called T2, was known to contain DNA surrounded by a protein coat.

Although aware of Avery's work, Hershey still believed that genes were encoded in protein, as did most scientists at the time. To test this notion, he exploited the newly available radioactive forms of phosphorus (^{32}P) and sulfur (^{35}S). Hershey and his research assistant Martha Chase prepared radioactive phage particles by infecting bacteria in medium that contained either isotope. Because DNA is rich in phosphate but not sulfur, and the opposite is true of proteins, these phage particles contained "labels" that were specific for each type of molecule. The radioactive phage particles were then used to infect fresh bacteria, and after the particles were allowed to attach to their host cells, the mixture was subjected to vigorous agitation using an ordinary kitchen item, called a Waring blender. The infected bacteria were then isolated from the blended mixture to determine where the labeled phage DNA and protein had gone. What Hershey and Chase found was both indisputable and, to them, quite astonishing. Most of the ^{35}S-labeled phage protein was shaken from the infected bacteria by blending. In contrast, almost all of the ^{32}P-labeled phage DNA was associated with the isolated bacteria. Furthermore, progeny phage particles produced by these bacteria contained ^{32}P, as would be expected for genetic inheritance via DNA. These results were exactly the opposite of what Hershey expected; he thought that the DNA "scaffold," would be washed away in the blender and much of the protein would be retained in the bacteria. Typically cautious in his interpretation, Hershey concluded that "the protein component(s) probably has no function in reproduction of the phage" but "the DNA does."[3] Nevertheless, this conceptually simple but elegant study converted all previous skeptics, and today, the Hershey-Chase experiment is among the most famous in scientific history. Later studies with a variety of phages and also some animal viruses showed that introduction of purified viral DNA into a host cell is all that is required for production of a new crop of progeny viruses.

Results from the "blender experiments" reached James (Jim) Watson, in a letter from Hershey in 1951. A precocious young member of the Phage Group, Watson had recently arrived in England as a postdoctoral fellow in the Cavendish Laboratory at Cambridge University, where he shared an office with Francis Crick. Watson had been interested in DNA as the source of genes for some time, and the news from Hershey fueled his growing passion to solve its structure. He was driven by the conviction that such knowledge would provide a key to the "secret of life." Believed by many to be "too bright to be really sound,"[4] the twenty-three-year-old Watson has been described most vividly by the eminent French biologist, François Jacob, as "an amazing character. A surprising mixture of awkwardness and shrewdness . . . childishness in the things of life and maturity in the things of science."[5]

Watson's thirty-five-year-old lab-mate, Crick, was an equally noteworthy character. He was described by Watson as having a quick, penetrating mind but never in "a modest mood" and "most people thought he talked too much."[6] Crick had been trained in mathematics and physics but was still working on his PhD research, which dealt not with DNA but with protein structure determination. Brash and unknown at the time, yet enticed by this most exciting challenge, Watson and Crick set out together to solve the structure of DNA.

The story of their success, as well as its momentous implications, has been told in numerous publications, including separate books authored by Watson and by Crick and even in TV movies. The two believed that that they could get the structure without doing any experiments by building a model of DNA from its known constituents, because the famous Caltech scientist Linus Pauling had recently been successful in solving protein structures this way. They were also urged on by the knowledge that Pauling was already working on models of DNA, and they did not want to be scooped by this established presence in the field. After a few false starts, their efforts were galvanized when Watson was shown the exquisite experimental data of Rosalind Franklin (although without her knowledge) by Maurice Wilkins, head of a group at King's College. Franklin's data, which aimed to derive DNA structure using X-ray crystallography, revealed DNA as a helical molecule made up of two strands. To construct their final model, Watson and Crick drew upon this insight, as well as an important correlation established by the biochemist, Erwin Chargaff, namely that the component four bases exist in complementary pairs in DNA: the amount of adenine (A) is always equal to the amount of thymine (T), and the amount of guanine (G) is always equal to the amount of cytosine (C). Their final model, built with brass plates and clamps, was a double helix formed by phosphate-linked sugar chains running in opposite directions and held together by chemical bonds between the complementary base pairs, much like the rungs on a spiral staircase.

In true British tradition of fair play, Wilkins was promptly invited to Cambridge to inspect the Watson-Crick model. He was immediately convinced of its credibility and excited by the implications. Two days later, Wilkins informed Watson and Crick that he and Franklin confirmed that their X-ray data strongly supported the double-helix structure. They were therefore anxious to publish the results of their physical analyses together with a report of the model. Relieved at this generous response, Watson and Crick readily agreed. Two papers from the King's College group were subsequently published together with the description of the Watson-Crick DNA model in the April 25, 1953, issue of the journal *Nature*.

The Watson-Crick DNA structure turned out to be enormously informative. The complementarity of bases in the two strands suggested that DNA could be duplicated via the separation and copying of each strand, thereby allowing genetic information to be passed on as cells divided. Furthermore, a possible alphabet comprising the four DNA bases could now be imagined, with "words" specifying particular amino

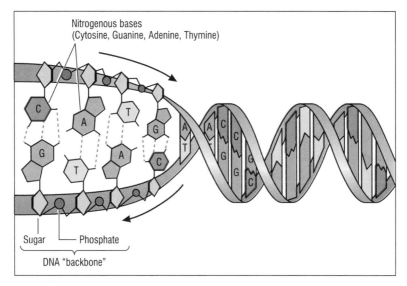

Nitrogenous bases
(Cytosine, Guanine, Adenine, Thymine)

C
A
T
G
A
G
A
T
A
C
C
G
G
G
G
C

Sugar —— Phosphate

DNA "backbone"

Fig. 1.3 The Watson-Crick double helix is held together by nucleotide base pairing. Bases are identified above the helix and components of the backbone are identified below.

acids encoded in their linear order. Elucidation of the structure of DNA marked the end of what has been called the "classical" period in genetics, a time when a few extraordinary scientists made momentous discoveries in what was still a relatively unknown field. In a sad footnote to history, Rosalind Franklin, a rather unwitting initial collaborator in this monumental work, died at age thirty-seven, just four years before the Nobel Prize was awarded to Watson, Crick, and Wilkins. As no more than three individuals can share a single Nobel Prize, we can only wonder how the honor would have been distributed if Franklin had lived.

The publication of the Watson-Crick model electrified the scientific community and revolutionized biology. The pace of discovery picked up substantially as numerous scientists in various fields, including biochemistry, physics, and genetics, raced to discover exactly how DNA is duplicated, how genetic information is encoded in DNA, and how such information can direct the production of the protein molecules determining the identity and properties of organisms. In fairly short order, Arthur Kornberg, a biochemist at Stanford University, isolated an enzyme from the bacterium *E. coli* that could make DNA. The newly discovered enzyme was able to join nucleotides into long polymers in the presence of a single strand of DNA, which served as a template. Consistent with Watson-Crick predictions, the assembly took directions from the template and the product contained a complementary sequence of bases, which ran in the opposite direction. This enzyme, a DNA-dependent DNA polymerase called Pol I, was the first of its kind to be identified. Many more nucleotide polymerases, including those encoded in viral genomes, would be discovered in the next few years. Meanwhile, in a famous and crucial confirmation of DNA structure, two young scientists, Matthew Meselson and Franklin Stahl, showed that DNA is duplicated in bacteria in a "semi-conservative" manner: the two strands are indeed unwound, and each daughter DNA duplex is composed of one conserved original strand and one strand that is newly synthesized.

The burning question that remained was, How can information in DNA direct the production of proteins? At that time, scientists knew that proteins are assembled on small dense particles in the cytoplasm, called ribosomes. These particles, first described in 1954, are composed of two subunits. Each subunit includes numerous proteins bound to a single, long strand of ribonucleic acid (RNA). RNA contains a sugar-phosphate backbone just like DNA; the principal differences between the two are that the sugar in RNA is ribose rather than deoxyribose, and uracil (U), rather than thymine (T), is one of the four bases. Other, as yet uncharacterized molecules of RNA were known at the time to be present in both the nuclei and cytoplasm of plant and animal cells.

The answer to the question of how DNA might direct protein synthesis was first hypothesized during an informal brainstorming session in 1960, which included the French geneticists François Jacob and Jacques Monod, Francis Crick, and the quick-witted, pixie-like South African, Sydney Brenner, in whose King's College rooms they and their colleagues had all gathered. Considering the available data from their work and that of others, the group agreed that DNA must direct protein synthesis via some sort of "messenger," which could travel in eukaryotic (plant and animal) cells from the nucleus to the cyto-

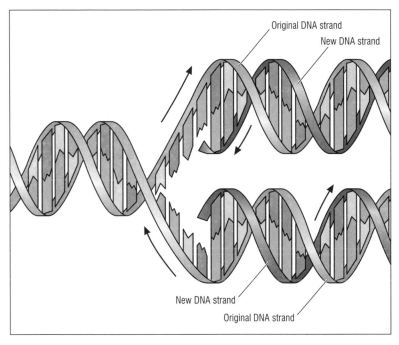

Fig. 1.4 Semi-conservative duplication of DNA. The duplex is unwound and each daughter molecule contains one completely conserved strand (gray) and a complementary strand (purple) that is newly formed by copying the conserved stand.

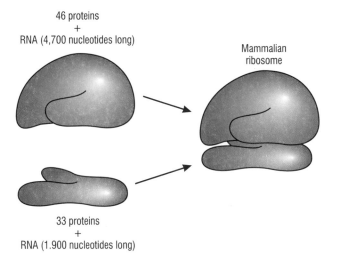

Fig. 1.5 Organization of the ribosome. Ribosomes comprise two subunits, each made up of numerous proteins bound to single long chains of RNA. The components of a mammalian ribosomes are listed.

plasm. Brenner let out a loud yelp when he realized that a molecule of RNA discovered in phage-infected bacterial cells two years earlier by Eliot Volkin and Lazarus Astrachan had exactly the right properties for such a messenger. The base composition of this RNA closely resembled that of the infecting phage's DNA. Furthermore, the RNA was short-lived, allowing protein synthesis to be "turned off" rapidly when needed. Subsequent experiments by Brenner, Jacob, Monod, and many others confirmed this prescient insight. A species of RNA, thereafter called messenger RNA (mRNA), which contains a copy of the sequence of a gene encoded in DNA, was established to be the intermediate in the transfer of information from genes for production of proteins by ribosomes. The process of gene expression in all living cells came to be understood in the context of two distinct steps. In the first step, called "transcription," a gene sequence in DNA is copied (via a DNA-dependent RNA polymerase) into mRNA. The mRNA is then delivered to ribosomes, where the second step occurs: "translation" of the mRNA sequence into the sequence of a specific protein.

In 1955, Francis Crick proposed that a kind of adapter molecule, one that carried an amino acid, would also be required to synthesize a polypeptide from a nucleotide code. That molecule was identified three years later as a small RNA, subsequently called transfer RNA (tRNA), by the biochemists Paul Zamecnik and Mahlon Hoagland. Later experiments showed that each tRNA molecule carries only one of the twenty amino acids found in proteins, attached to its flexible tail. Once mRNA was identified, it was recognized that base pairing between particular sequences in mRNA and their complementary tRNAs could be translated into the sequence of a protein when these RNAs were brought together in the "reading head" of a ribosome.

Although this general scheme for information transfer from gene to protein through RNA intermediates was accepted by the early 1960s, the "genetic code" (i.e., the number and sequence of bases in mRNA that specified each amino acid) remained a mystery. The challenge of solving this puzzle was taken on by a number of researchers, using quite different approaches. Based on mathematical considerations, the Russian theoretical physicist George Gamow proposed a code comprising

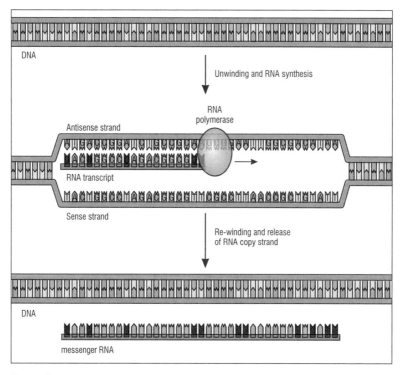

Fig. 1.6 Transcription of messenger RNA from genes in DNA. When a gene is to be expressed, the DNA strands separate such that one of them, called the antisense strand, serves as a template for formation of a messenger RNA (mRNA) containing information corresponding to that in the sense strand of DNA. RNA synthesis is mediated by a DNA-directed RNA polymerase in association with other proteins. The mRNAs are then transported to ribosomes, where they direct synthesis of the specified proteins.

three bases as a minimal requirement, because a two-base code would be sufficient to specify only sixteen of the existing twenty amino acids.

In 1954, Gamow recruited a group of prominent scientists (mostly his friends and *all* male) into a scientific "gentlemen's" club, called the "RNA Tie Club." The dual mandate of the club was to solve the structure of RNA and decipher the genetic code. Each of the twenty-four members was given a woolen tie with a green and yellow helix and a pin with the name of one of the amino acids. Four honorary members were given the names of each of the RNA bases. This brotherhood included theoreticians, physicists, chemists, and biologists, some of whom had no obvious connection to the stated goals of the club. However, six of the members were or would become Nobel laureates: the chemist Melvin Calvin (code name, histidine), theoretical physicist Richard Feynman (glycine), Watson and Crick (proline and tyrosine), Max Delbrück (tryptophan), and Sydney Brenner (valine). As it turned out, the club never met as a whole, but "local chapters" had occasional get-togethers (with cigars, alcohol, and associated camaraderie), and members circulated ideas and manuscripts for comments, including Crick's adapter hypothesis described previously. Although Crick's genetic experiments suggested that a sequence of three bases in mRNA (called a codon) would comprise information specific for one amino acid, in the end, neither RNA structure nor the genetic code was solved by this illustrious group. Rather, it was through the methodical "wet-lab" research of nonmember biochemists that the code was finally cracked.

Marshall Nirenberg, a newly established investigator at the U.S. National Institutes of Health, paved the way to deciphering the code by developing a test-tube system that could interrogate RNA sequences directly. His approach was inspired by an earlier report that polymers (called polypeptides) composed of a single amino acid, phenylalanine, were formed when an RNA polymer that contained only U was incubated with ribosomes. This result implied that some multiple of U was the codon for phenylalanine (Phe). Using bacterial (*E. coli*) ribosomes and tRNA that carried radioactive phenylalanine, Nirenberg discovered that the minimal length of polyU required for radioactive tRNA^{Phe} to be bound to the ribosomes was three: UUU. With this crucial experiment, Nirenberg had obtained the very first direct evidence that the language of genes is actually written in words of three letters.

Nirenberg first presented his results at a biochemistry conference in Moscow. Those present immediately recognized that all sixty-four possible combinations of three of the four nucleotides (i.e., $4 \times 4 \times 4$) could be decoded in this way. The scientific community was galvanized and a race was on to identify the remaining codons.

The pace of discovery accelerated when Gobind Khorana at the University of Wisconsin devised a method to synthesize RNA polymers

Second Postition

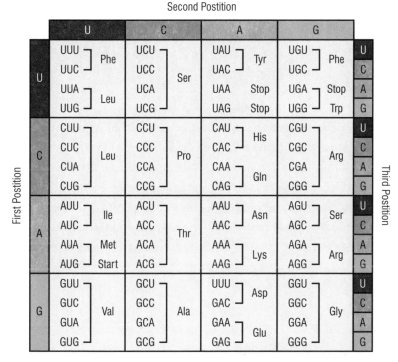

Fig. 1.7 The genetic code. The three-letter words, called codons, are read from right to left. The language is universal; with very few exceptions, every living thing uses the same codons. Because more than one three-letter codon can specify a particular amino acid, the code is considered "degenerate." For example, either UAU or UAC can direct incorporation of the amino acid tyrosine (Tyr) into a protein, whereas any one of three codons (UAA, UAG, or UGA) can cause protein synthesis to terminate. The codon AUG specifies the amino acid methionine (Met) and the start of protein synthesis.

having a defined base sequence. The first sequence he made and tested was a polymer with alternating U and C bases (UCUCUC . . .), which could therefore contain two potential, nonoverlapping three-letter codons. Khorana found this polymer was translated into a polypeptide comprising repeats of serine followed by leucine, thereby defining the codons for these two amino acids as UCU and CUC, respectively. Synthetic RNA polymers were later used to decipher many other "words" in the genetic code, including those that specified the first and the last amino acid in a protein. Amazingly, by 1961, only eight years after the solution of DNA structure, a method for cracking the genetic code was at hand. Nirenberg and Khorana were awarded a Nobel Prize for this work in 1968.

The Stage Is Set

Just as retroviruses were being discovered as agents of disease, the fields of genetics and molecular biology had expanded enormously and reached a historic milestone. A broadly applicable and generally accepted understanding of the chemical and structural nature of genetic material was at hand, as well as the way in which information encoded in DNA could be passed through to the next generation. It was also understood that several types of RNA molecules are needed to translate the information encoded in genes to produce proteins: mRNAs as gene

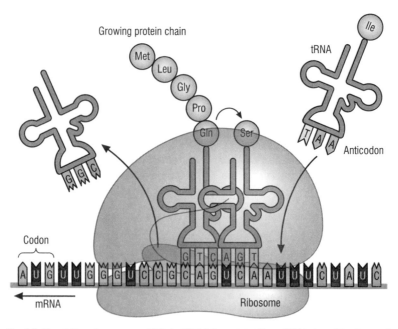

Fig. 1.8 Translation of messenger RNA (mRNA) into protein. The mRNA is bound by the small ribosomal subunit and fed in toward the left, while amino acids bound to transfer RNAs (tRNAs) are joined in the large ribosomal subunit. The protein sequence is specified by base paring between codons in the mRNA with anticodons in the tRNAs and proceeds in a linear fashion.

copies, tRNAs that contain codon complements along with specific amino acids, and RNA-containing ribosomes.

The route for transmission of genetic information came to be expressed in universally accepted shorthand as

$$DNA \rightarrow RNA \rightarrow protein.$$

This unidirectional pathway was to become the "central dogma" of biology. It seemed at that time that a great principle of biology had been established and only the molecular and biochemical details of the accepted dogma needed to be defined. This task was eagerly embraced by many scientists. Viruses became especially popular tools for further research in genetics and molecular biology because their genomes were small and initiated a defined cycle of information transmission upon infection of host cells. Little did anyone suspect that elucidation of the unique replication cycle of retroviruses would shatter the central dogma.

Amending the Central Dogma

Because viruses depend on their host cells for reproduction, study of their genomes and the manner in which cellular functions are coopted during their reproduction can reveal details of the biology of both the virus and its host. By the mid-1990s, it was known that DNA is the heredity component in all living cells and in many viruses. However, some viruses, especially those that infect plants, were found to contain RNA but no DNA. This was the case for tobacco mosaic virus, one of the first viruses to be discovered. But the fact that RNA comprised the genetic material of this plant virus did not become apparent until 1955, through a series of experiments that began by the truly remarkable discovery that infectious particles of the tobacco mosaic virus could be reconstituted from their purified components.

Heinz Fraenkel-Conrat and his colleagues at the University of California, Berkeley found that by mixing clear solutions of the chemically separated tobacco mosaic viral protein and viral RNA and simply waiting for twenty-four hours, they could see particles in the electron microscope that looked just like the original virus. Although we now know that tobacco mosaic virus is unique in its capacity to assemble spontaneously, the fundamental simplicity of these agents was underscored dramatically by these remarkable observations.

However, like other researchers at the time, these investigators assumed that the proteins of tobacco mosaic virus comprised its hereditary component. To test this idea, they mixed purified viral proteins and viral RNAs from different strains of the tobacco mosaic virus and then infected tobacco plants with these recombined, reconstituted particles. Their analysis of virus progeny produced in the infected plants gave a result that they did not expect: the protein composition in progeny particles depended on the origin of the RNA in the reconstituted particles. It was apparent, therefore, that RNA, not protein, must be the genetic component of these viruses. Their experiments showed clearly that while the viral protein is required to form an infectious particle, it is the nucleic acid component that contains the genetic information for formation of this protein. Furthermore, the finding that the genome of tobacco mosaic virus comprises RNA was recognized as a major discovery at the time. It would soon become apparent that RNA genomes are not restricted to the viruses of plants.

The first bacterial virus with an RNA genome was discovered serendipitously in the late 1950s by Tim Loeb, a graduate student of Norton Zinder at the Rockefeller University. Loeb was hunting for new bacteriophages that could infect a certain strain of the bacterium *E. coli*. As sewage is the most likely place to find phages of a human gut bacterium, he visited a local sewage plant in New York City and obtained a large, heterogeneous sample to screen. Loeb's efforts were rewarded with the isolation of two new phages. Chemical analyses showed that one of the isolates had a DNA genome, similar to other known *E. coli* phages. But, quite unexpectedly, the other isolate contained RNA but no DNA. Later studies showed that the RNA genomes of the tobacco mosaic virus and a bacteriophage related to Loeb's isolate could both serve directly as messenger RNAs (mRNAs) for the production of proteins. Although the existence of RNA genomes in some viruses required modification of the central dogma, these results were nevertheless consistent with the expected unidirectional flow of information, RNA → protein. Unlike host cells and viruses with DNA genomes, it appeared that the reproduction of "RNA viruses" could proceed via a shortcut that eliminated the need for DNA.

Many other examples of viruses with either DNA or RNA genomes were discovered by the 1960s. (Indeed, new viruses continue to be discovered to this day.) Those most relevant to our story, however, are the viruses that promoted tumor formation in animals. These agents were divided into two general categories, depending on the nature of their genomes: the DNA tumor viruses, such as the polyoma virus of mice, and the RNA tumor viruses, which included Rous sarcoma virus and other retroviruses known to cause malignancies in birds and mammals. The identification of tumor viruses sparked intense interest among scientists and the public. In 1964, after a flurry of publicity about viruses and cancer, a Virus Cancer Program was created at the U.S. National Cancer Institute to promote research on these viruses. With this additional support and the shared resources made available through the program, many scientists entered the field. However, progress was initially slow, primarily because quantitative methods and convenient cell culture systems were not yet available for these viruses.

A New Way to Study Animal Viruses

Many of the early advances in genetics, including elucidation of the central dogma, were based on knowledge gained from research with bacteria and their viruses, the bacteriophages. Such studies benefited from the ease with which bacterial cells could be cultured in the laboratory and the quantitative methods that had been developed to study bacteriophages and their genetics. For example, the concentration of infectious bacteriophages in a preparation could be determined by a relatively simple method called a plaque assay, which had been standardized in 1943 by Alfred Hershey and was used extensively by the Phage Group. In this assay, samples from serial dilutions of a phage preparation are mixed with an excess of host bacteria. The mixture is then added to a semisolid solution of agar (a gel-like substance made from seaweed) that is spread over a nutrient rich layer of solid agar in a petri dish. The bacteria multiply and produce a "lawn" of cells in the semisoft top layer, except in the spot where a bacterium has been infected with a phage particle. Within an hour after infection, this cell will explode (lyse) on the lawn, releasing about a hundred new phage particles. These particles can then infect neighboring bacteria but cannot go much further because diffusion is restricted in the semisoft agar. The net result is a sort of circular chain reaction of infections that lead to bacterial cell lysis, which creates a visible hole (a plaque) in the bacterial lawn within twenty-four hours. From the number of plaques on the plate, one can determine the number of infectious phage particles that were present in the sample and, by back calculation, the concentration in the original preparation. Variations in the sensitivity of different host strains to infection and differences in the size and clarity of

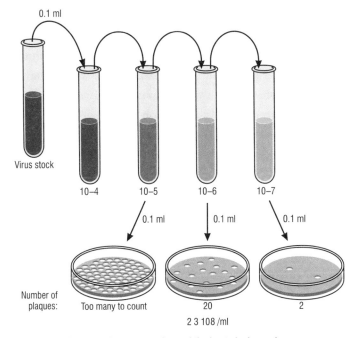

0.1 ml

Virus stock

10–4   10–5   10–6   10–7

0.1 ml   0.1 ml   0.1 ml

Number of
plaques: Too many to count   20   2

2 3 108 /ml

Fig. 2.1 Serial dilutions of a virus preparation and the bacteriophage plaque assay.

plaques made it possible to isolate a variety of phage mutants and to study the genetic basis for their distinct properties.

Application of the plaque assay was central to the successful analysis of bacteriophage genetics by members of the Phage Group. Alfred Hershey, Salvador Luria, and the physicist Max Delbrück at Caltech, the three founding members of the group, were each extraordinary scientists with unique personalities contributing to the "Church," as the group was sometimes called.[1] In this context, Hershey, the taciturn but superlative experimentalist, was known as the "saint," Luria's social conscience and outgoing personality earned him the label of "priest / confessor," and Delbrück's "*ex cathedra*" manner made him the "pope." The three shared the 1969 Nobel Prize in Physiology or Medicine "for their discoveries concerning the replication mechanism and the genetic structure of viruses." Their work established the foundation for the field of molecular biology, the study of how molecules behave in living organisms.

In contrast to the bacteriophages, research with animal viruses was initially cumbersome and inexact. Concentrations of infectious animal virus particles were ordinarily estimated by injecting serial dilutions into susceptible animals or chicken embryos and determining the dilution at which infection was lost. These methods were both time-consuming and costly, making it difficult to conduct genetic studies with these agents. As methods for culturing some animal cells in the laboratory improved, it seemed possible that the use of such cells could simplify animal virus quantification and genetic analyses. The challenge of using cultured cells to develop a plaque assay for animal viruses was taken up by the Italian physician and researcher, Renato Dulbecco.

Dulbecco was among the many young European scientists whose careers were put on "hold" by World War II. He had been a student in Italy with Salvador Luria but, unlike his friend, did not leave that country at the start of the war. Dulbecco was conscripted into the Italian army as a medical officer to serve in France and on the Russian front. Seriously wounded during a major Russian offensive, he was shipped back to Italy to recover and later joined the resistance, serving as a physician. After the war, Dulbecco attempted to reestablish his career while working in his previous mentor's group in Turin, Italy. However, conditions in postwar Italy made research extremely difficult. Consequently, when Luria invited Dulbecco to come to his laboratory in the United States during a visit to Turin, Dulbecco was ready to leave.

Dulbecco's work as a virologist began in 1947 as a research associate in Luria's group at Indiana University, where he was soon joined by Luria's graduate student, James Watson (later of DNA fame). Dulbecco learned the quantitative methods established for bacteriophages and showed that exposure to visible light could reverse the lethal effects of UV light on these phages, a phenomenon dubbed "photo-reactivation." Having proven his research competence, Dulbecco was invited to join Delbrück's group at Caltech. After working at Caltech with phages for a few years, Dulbecco was asked by Delbrück if he would be interested in studying animal viruses. A wealthy donor had provided a fund to start a new research program with animal viruses—in which Delbrück had no interest. Dulbecco readily agreed to take on this project; as a physician-scientist, animal virus pathogenesis was of interest to him. Delbrück arranged for one of his research fellows, Marguerite Vogt, to join Dulbecco in this new project. Vogt was also a European émigré, with an MD degree from the University of Berlin. Although she was working on bacteriophages with Delbrück, Vogt had performed important research on the genetics and development in fruit flies while in Germany. Her formidable talents and skill in cell culture were to prove most valuable to Dulbecco, with whom she was a long-term collaborator. Vogt was among several pioneering women scientists whose talents were acknowledged, but advancement was stunted by barriers known today as the "glass ceiling."

Dulbecco realized that genetic analyses of animal viruses would be greatly facilitated if he could develop quantitative methods similar to those he had used with bacteriophages but with cultured chicken or mammalian cells as hosts. His first major accomplishment as an animal virologist was the creation, together with Vogt, of a plaque assay for animal viruses that kill their host cells during propagation. This assay uses single layers of cultured animal cells that can be grown on the bottom surface of petri dishes. To allow cells to be infected, the monolayer is exposed to samples of a virus for an hour or so. The cells are then covered with nutrient-containing agar, to keep any virus particles produced from the initially infected cell from diffusing too far. After a day or two, the agar layer is removed and the living cells in the monolayer are stained with a dye, revealing holes (plaques) in locations where the initially infected cell and its neighbors had been killed by the virus. Dulbecco's plaque assay, published in 1952, was widely adopted by other animal virologists and thereafter set the standard for quantitative studies with many animal viruses. Dulbecco's laboratory went on to make seminal contributions to understanding the reproductive cycles of animal viruses that kill (lyse) their host cells.

The Rous sarcoma virus was of major interest to Harry Rubin, a veterinarian who joined Dulbecco's group in the late 1950s. To devise a quantitative assay for this virus, Rubin enlisted the help of Howard Temin, a gifted first-year graduate student of Dulbecco. They exploited the fact that while this virus does not lyse its host cell, it does induce the cell to proliferate aberrantly, a process called transformation. They developed an assay that starts with establishment of a sparse monolayer

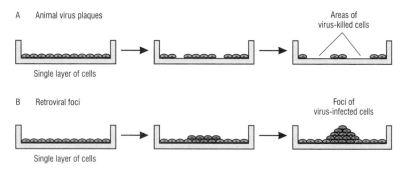

Fig. 2.2 Plate assays for animal viruses. (A) Plaques for lytic animal viruses comprise "holes" in the monolayer of cultured cells. (B) Foci for oncogenic retroviruses like the avian Rous sarcoma virus comprise multilayered piles of cells.

of chicken embryo cells. After the cells are exposed to appropriate dilutions of the virus, they are covered with soft agar, as in Dulbecco's plaque assay. Normally, when uninfected cells in a growing monolayer become so numerous that they begin to touch each other, they stop multiplying and lay flat on the culture dish, a phenomenon known as contact inhibition. Because a cell that is infected with the Rous sarcoma virus lacks such inhibition, it will continue to multiply, eventually forming clumps of transformed cells in the monolayer, called foci. As with Dulbecco's plaques, each focus identifies the spot in which a single host cell has been infected. In a landmark paper published in 1958, Temin and Rubin described their assay, which remains a standard in the field to this day.[2] Their focus assay was the starting point for uncovering many of the other unique properties of these RNA tumor viruses.

The DNA Provirus Hypothesis

Temin stayed on in Dulbecco's laboratory for a short time after receiving his PhD degree but then established his own laboratory at the University of Wisconsin in 1960, where he continued to study the Rous sarcoma virus. He had become intrigued by the fact that the foci formed by diverse strains of the virus looked different. For example, some foci were made up of fat, rounded cells, whereas others contained long, spindle-shaped cells. Furthermore, when viruses produced by cells from each type of focus were assayed, the cells they infected formed foci identical to those from which the viruses had been derived. Temin concluded that the shape (morphology) of the transformed cells must be dictated by genetic information in the virus. He became convinced, therefore, that the viral genome must be maintained in the infected cell in some sort of stable form.

Aspects of the behavior of the Rous sarcoma virus were comparable to certain bacteriophages studied by the Phage Group. These bacteriophages are called "temperate" because they do not always kill their host cells, especially if conditions in the cell are unfavorable for phage propagation. Instead, the DNA from the infecting phage is spliced into the genome of its host bacterium, where it is copied together with the host DNA as the cell grows and multiplies. The phage DNA becomes a part of the cell's inheritance through future generations. The inserted phage DNA is called a "prophage," and its host bacterium is considered "lysogenic" because the prophage can occasionally be induced to initiate a cycle of progeny virus production, which then leads to lysis of the cell.

Temin realized that there were important differences between the temperate phages and his Rous sarcoma viruses. Lysogenic bacteria do not produce progeny phage until the prophage is reactivated by some change in conditions within their host cell. In contrast, transformed chicken cells produce numerous progeny Rous sarcoma virus particles continuously, and they do so without being killed. Most important, the genome of the Rous sarcoma virus comprises RNA, not DNA. Nevertheless, Temin made a conceptual leap based on his genetic analyses of transformed chicken cells, as well as other experiments, and presented his radical ideas at the 1964 International Conference on Avian Tumor Viruses: Temin hypothesized that the RNA genome of the infecting virus somehow acts as a template for the synthesis of viral DNA, which he called the "provirus" in analogy to the bacterial prophage. Furthermore, he proposed the provirus persists as an integral component of the genetic inheritance of its host cell, where it provides a template for synthesis of progeny virus.

At the time, scientists were aware of enzymes that copied DNA from DNA templates and those that used DNA to make RNA. But there was no known mechanism by which RNA could be copied into DNA. Although the results of Temin's experiments yielded data consistent with his hypothesis, his findings were also open to alternative interpretations, as the techniques available at the time were not up to the task. Furthermore, the unidirectionality of the central dogma was etched in the collective scientific consciousness: DNA $\rightarrow$ RNA $\rightarrow$ protein. For most of his contemporaries, the implied reversal of information flow enshrined in the dogma was just too heretical to be accepted. Consequently, Temin's DNA provirus hypothesis was disbelieved by most, scorned by some, and essentially disregarded for the next six years.

Discouraging as this might seem, Temin did not abandon his ideas. Although no single experiment he performed was entirely conclusive, the results were consistent with his hypothesis. Firm in his own convictions and truly courageous in the face of widespread incredulity, he continued to look for more compelling evidence. As it happened, Temin was also not entirely alone in his quests.

An independent series of experiments with Rous sarcoma virus, conducted in Prague, had also persuaded Jan Svoboda that there must be a DNA form of the viral genome. Svoboda discovered that a particular strain of Rous sarcoma virus produced tumors when injected into rats. While the cell lines that he derived from such tumors did not produce virus, they maintained their transformed morphology through many generations. Furthermore, infectious Rous sarcoma virus particles were produced when the transformed rat cells were injected into chickens. Because of this result, Svoboda considered the rat cells to be "virogenic," by analogy with lysogenic bacteria. Svoboda independently proposed that the virogenic rat cells contain a DNA provirus in a 1963 publication of Charles University in Prague. Temin and Svoboda corresponded during the period from 1962–1963, exchanging thoughts and even some materials, as they each devised experiments to convince their skeptical contemporaries of the validity of their hypothesis. For Temin, the critical breakthrough came in 1969 following the arrival in his laboratory of the young Japanese biochemist, Satoshi Mizutani, who would search for an enzyme capable of making DNA from the viral RNA genome.

Viral Enzymes and the Discovery of Reverse Transcriptase

David Baltimore began his graduate studies at the Massachusetts Institute of Technology (MIT) working with phage in the laboratory of Salvador Luria, who remained a lifelong friend and advisor. With Luria's blessing, Baltimore transferred to the Rockefeller University in the early 1960s after becoming fascinated by a fundamental question raised with the discovery of viruses with RNA genomes: all cellular RNAs are produced from a DNA template, so how are the genomes of RNA viruses replicated? As a student at the Rockefeller, Baltimore focused his attention on Mengovirus, a mouse virus in the same family as poliovirus, named for a district near Kampala, Uganda, where it was first discovered. Like tobacco mosaic virus and the RNA phage described earlier, this family of viruses contains single-strand RNA genomes capable of directing protein synthesis, analogous to cellular mRNAs. Baltimore and his mentor, Richard Franklin, showed that Mengoviral RNA synthesis takes place in the cytoplasm of its host cell, not in the nucleus where DNA directs the synthesis of all cellular RNAs. They also found that an inhibitor that binds to DNA in the nucleus and blocks the synthesis of cellular RNA had no effect on viral RNA synthesis. Therefore, it appeared that the viral RNA genome must encode information for the production of its own, viral replicating enzyme. By adding ribonucleotide substrates to a fraction isolated from the cytoplasm of infected cells, Baltimore and Franklin were able to detect viral RNA polymerizing activity. The 1962 report of their findings provided the first evidence that an RNA virus genome encodes its own RNA "replicase."[3] Similar reports for RNA phages and other animal RNA viruses were to follow.

After completing his PhD research at the Rockefeller, Baltimore spent short intervals as a postdoctoral fellow back at MIT in Boston and at the Albert Einstein College of Medicine in New York. While at the latter, he acquired a variety of additional skills, including further expertise in nucleic acid biochemistry from Luria's accomplished brother-in-law, Jerard Hurwitz, co-discoverer of the cellular DNA-dependent RNA polymerase. Baltimore continued to work with RNA viruses throughout this period and was subsequently invited to set up an independent research program by Renato Dulbecco, who had just moved to the Salk Institute from Caltech. After a little over three years and having established a credible record of research focused on poliovirus replication, Baltimore returned to MIT in 1968 as a junior member of the faculty.

This was a truly exciting time for a young investigator interested in virus biochemistry. Two laboratories had just reported that nucleic acid polymerizing enzymes could be found *inside* of virus particles, a place where no one expected anything more than a viral genome to reside. A DNA-dependent RNA polymerase was identified in rabbitpox virus, a large DNA virus related to the virus that causes smallpox, and an RNA-dependent RNA polymerase was found in a virus with a double-strand RNA genome, reovirus. Baltimore began to wonder if the RNA virus that his wife and collaborator, Alice Huang, had studied earlier might also contain such an enzyme. This virus, vesicular stomatitis virus (VSV), harbors a single-strand RNA genome that cannot serve as mRNA for the production of viral proteins. Because there is no enzyme within the cell that can make mRNA from RNA, it seemed possible that VSV particles might contain a viral enzyme to do this job. In short order, Huang, Baltimore, and their colleagues did, indeed, identify an RNA-dependent RNA polymerase in VSV and in another virus with a similar RNA genome. Baltimore then considered Temin's radical hypothesis concerning the RNA tumor viruses.

David Baltimore was four years younger than Howard Temin, but their paths had crossed several times. As a high school student, Baltimore participated in a summer science program at the Jackson Laboratory in Bar Harbor, Maine, where Temin was a mentor for the new students. The two became friends, with Baltimore looking to the already accomplished Temin as a role model. Baltimore attended Swarthmore College, where Temin had been a somewhat rebellious but outstanding student, predicted to be one of the future giants in experimental biology. Moreover, both had been associated with Renato Dul-becco at formative times in their careers. Baltimore was well aware that the scientific community had rejected the DNA provirus hypothesis, but knowing Temin, he was not as ready as most of his colleagues to dismiss Temin's heretical ideas. He decided in 1970 to test the notion that RNA tumor virus particles contain an enzyme that can use RNA as a template for the synthesis of DNA.

Baltimore obtained a preparation of an RNA tumor virus of mice, the Rauscher murine leukemia virus, from colleagues at the National Institutes of Health and incubated the particles in a mixture that in-cluded the four substrate deoxynucleotides. Although the activity was barely detectable at first, he found that the added substrates were in-corporated into a polymer of DNA. Furthermore, DNA synthesis was abolished if the particles were first treated with an enzyme that de-stroyed the viral RNA. Baltimore then tested a preparation of Rous sarcoma virus particles and detected the same RNA-dependent DNA polymerase activity. With great exhilaration, he realized that here, finally, was concrete evidence to support the DNA provirus hypothesis. He quickly sent a manuscript describing his sensational results to the prestigious scientific journal *Nature,* before anyone might beat him to the discovery. Then he called his friend Temin to share his news.

What Baltimore did not know was that Satoshi Mizutani had ob-tained the very same result in Temin's laboratory using particles of the Rous sarcoma virus. Like Baltimore, Mizutani had been motivated by the discoveries of enzymes in other virus particles and also by Huang's and Baltimore's report of an RNA-dependent RNA polymerase activity in VSV. In their telephone conversation, Temin informed Bal-timore that while Mizutani's work had not yet been written up, he had publicly announced their findings just a few days earlier at a conference in Houston, Texas. After recovering from the shock of simultaneous discovery, the two scientists quickly contacted *Nature* and arranged for Temin to submit his manuscript with Mizutani. The reports from the Baltimore and Temin laboratories were considered so momentous that their papers were published back-to-back in the June 2, 1970, issue, just a few weeks after Temin's submission. Because the newly discovered viral enzyme *reversed* the direction of information flow specified by the central dogma (i.e. RNA → DNA, rather than DNA → RNA), the journal editors coined a new name for it: "reverse transcriptase." The RNA tumor viruses eventually came to be known as "retroviruses" (*retro* = backward in Latin). In recognition of their pivotal and paradigm-shifting research, Temin, Baltimore, and Dulbecco were awarded the 1975 Nobel Prize in Physiology and Medicine "for their discoveries concerning the interaction between tumor viruses and the genetic material of the cell."

Detection of the reverse transcriptase in retrovirus particles showed clearly how retroviral DNA was made, but it did not provide direct proof for the provirus hypothesis. The idea that retroviral DNA, embedded in the genome of its host cell, can direct synthesis of progeny particles was confirmed in 1972 by the husband and wife team, Miroslav Hill and Jana Hillova. These investigators treated chicken cells with purified DNA from the rat cell line derived from Svoboda's Rous sarcoma virus–induced tumors. They found that the treated chicken cells then produced retrovirus particles that were identical to the Rous sarcoma virus that Svoboda had used to induce the rat tumors. Hill and Hillova concluded that even though progeny virus particles were not made in the rat cells, proviral DNA had been maintained as an integral component of the DNA of these cells through many generations. All that was needed for the provirus to direct production of progeny particles was the congenial environment of its normal avian host cell.

The discovery of reverse transcriptase, proof of the existence of a provirus, and the potential use of these viruses to study their effects on their host genomes encouraged many researchers to enter the retrovirus field. The application of DNA cloning, sequencing technologies, and other new methods led to a burst of activity and valuable new insights. By the 1980s, it was apparent that viral DNA synthesis occurs rapidly after a retrovirus enters its host cell and the accompanying reverse transcriptase gains access to a supply of substrate deoxynucleotides. By analysis of viral DNA produced and integrated in infected mouse and chicken cells, it was determined that the following features are shared among all retroviruses: the major final product of reverse transcription is a linear, double-stranded DNA copy of the viral RNA genome, with one unique distinction—the DNA copy contains long repeats at each end called LTRs (for long terminal repeats), made up of sequences derived from both ends of the viral RNA. Subsequent studies showed the LTRs contain signals that control the synthesis of mRNA after the viral DNA is inserted into the host cell's genome. Short duplications of host DNA that flank each LTR of the integrated viral DNA copy (the provirus) was another shared feature, shown in the early 1980s to be produced during the process of retroviral DNA integration.

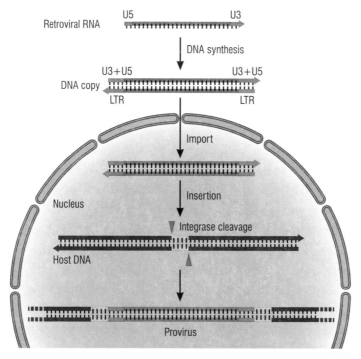

Fig. 2.3 Retroviral DNA synthesis and its integration into host DNA. Retroviral RNA is depicted in green and DNA in gray. Sequences unique to either end of the viral genome (U5 and U3) are colored orange, and host cell DNA is black. During integration, two base pairs are removed from one strand of viral DNA at each end of the LTR (indicated by the arrow tips), and a staggered cut is made in a target sequence in the host DNA, which is four to six base pairs long, depending on the virus. In the figure, a target sequence is illustrated in yellow and the cleavage sites with vertical red arrows. Following attachment of the cut ends of the viral DNA to the cleaved host DNA, repair of the staggered cuts results in a duplication of the host target sequence on either side of the provirus.

The ability to isolate reverse transcriptase (abbreviated RT) from retrovirus particles propagated in chicken or mouse cell lines encouraged numerous laboratories to study the process of reverse transcription in great detail. The results revealed an amazingly complex but elegant mechanism in which the viral RNA genome is destroyed as it is copied into a single DNA strand. The second, complementary DNA strand is then synthesized by the reverse transcriptase, with sequences at the ends being copied twice to form the LTRs.

Retroviral Particles Contain Not One but Three Enzymes

Although a general outline of how reverse transcriptase produces viral DNA was acquired rapidly in the 1980s, the question of how this DNA is inserted into the host genome remained a mystery. But once again, there were clues to be had from the bacteriophages. Some temperate phages were known to encode a special recombination enzyme, called integrase, which is able to splice the viral DNA into the genome of its bacterial host. Therefore, it seemed possible that retroviruses might possess a similar enzyme. Furthermore, if the reverse transcriptase is sequestered in retroviral particles, an integrase protein might also be there. Reasoning that such an enzyme would need to bind DNA, biochemist Duane Grandgennet first looked in disrupted retroviral particles for a protein with this capability. After finding such a small protein, distinct from reverse transcriptase, he showed that it could also cut DNA. Such an activity was consistent with the notion that host DNA must be cleaved in order for the viral DNA to be inserted. Grandgennet's 1978 report of this work was the first identification of what he and others would later confirm to be the retroviral integrase (abbreviated IN).[4] Subsequent work revealed that integrase proteins recognize and bind to the tips of the two LTRs in viral DNA and insert both ends into host DNA.

Reverse transcriptase and integrase are not, however, the only enzymes encased in all retroviral particles. A third viral enzyme, called protease, was identified in Rous sarcoma virus particles in the late 1970s.

The retroviral protease (abbreviated PR) also plays an essential role in the virus life cycle. The protein components of retroviruses, like those of many other animal viruses, are produced in infected cells as long precursor proteins in which the enzymes and structural components are joined end to end, like linked sausages. This process makes sense, as it ensures that all of the necessary constituents will enter a nascent particle in proper order and amounts. The job of retroviral PR is to snip at the links in the precursors so that the individual enzymes and structural proteins (by analogy, the sausages) are liberated at the right time following particle assembly. Because each of the three enzymes found in all retroviruses is essential for virus reproduction, they have been the major targets for antiviral drug discovery in the fight against AIDS.

Retrovirus Architecture and Reproduction in an Infected Cell

Invention of the electron microscope and later improvements in this technology made an enormous impact on the study of viruses. Its successful application by Wilhelm Bernhard in France and Joseph and Dorothy Beard at Duke University produced the first clear views of retrovirus morphology in 1958. Their images of retrovirus particles isolated from the blood of leukemic chickens, or visualized in thin slices through infected cells, revealed spherical structures containing a well-defined inner core surrounded by a membrane. The composition of these particles and details of their assembly in infected cells were the subjects of intensive study in the following decades.

The inner cores of retroviruses were found by biochemical and biophysical studies to comprise a protective shell, called a capsid (from the Latin word for box), formed by numerous interlocking copies of a single protein (designated CA) that surround the viral RNA and enzymes (PR, RT, and IN). Copies of a small basic protein, the nucleocapsid (NC), coat the viral RNA in the capsid. Two additional proteins contribute to the structure of retrovirus particles. Molecules of one, called matrix protein (MA), line the area beneath a lipid membrane derived

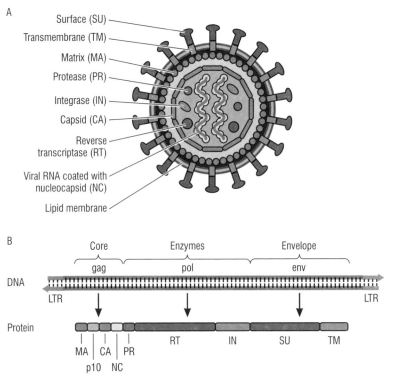

A

Surface (SU)
Transmembrane (TM)
Matrix (MA)
Protease (PR)
Integrase (IN)
Capsid (CA)
Reverse transcriptase (RT)
Viral RNA coated with nucleocapsid (NC)
Lipid membrane

B

Core — gag
Enzymes — pol
Envelope — env

DNA

LTR   LTR

Protein

MA | CA | PR   RT   IN   SU   TM
p10   NC

Fig. 2.4 Particle architecture and genome organization of the prototype avian leukosis virus. (A) Diagrammatic cross section of the virus particle indicating the location of components. (B) Genome map. The location of sequences encoding the viral proteins is indicated below the *gag, pol,* and *env* genes.

from the host cell, known as the envelope, which surrounds the capsid. Protrusions from the envelope that were seen in the electron microscope on the outside of the envelope, called "spikes," are composed of the second viral protein. The surface portion (SU) of the spikes contacts the matrix via a transmembrane portion (TM).

One very curious finding was that each retrovirus capsid contains not one but two viral RNA genomes. This feature is all the more mysterious because genetic analyses showed that viral DNA can be made from a single RNA genome and, furthermore, that only one viral DNA copy is normally made from an infecting particle. Although we still don't know why the encasement of two genomes was selected during retrovirus evolution, we do know that single-strand RNA is extremely sensitive to damage by various environmental assaults, such as UV light. Consequently, one idea generally accepted across the scientific community is that the packaging of two viral genomes is a form of genetic insurance. Even if both genomes are damaged or broken, a good DNA copy can still be produced by the reverse transcription of undamaged bits of either RNA genome.

Approximate maps for the location of genes on the retroviral genome were first obtained by genetic crosses and analysis of viral RNA fragments. However, the development of DNA cloning and sequencing methods in the late 1970s made it possible to derive precise maps simply by comparing the coding sequences in DNA copies of the retroviral genome with the amino acid content at the ends of the viral proteins. The prototype avian Rous sarcoma virus genome was one of the very first to be defined completely: sequences that encode proteins that make up the viral core are located at the beginning of the genome, adjacent to the "upstream" LTR. Sequences that encode enzymes are in the center, and those encoding the envelope spike protein are at the end of the genome. These genes and their order are common to all infectious retroviruses, although some groups of retroviruses have more complex genomes with sequences that encode additional proteins, generally located on either side of the genes for the viral core or envelope proteins.

Viruses are essentially inert and passive agents outside of their host cell, and their propagation depends entirely on the probability that one or more of the many progeny produced from one infected cell will find itself in the vicinity of another potential host. A retroviral infection begins when the spike proteins on the envelope of the retroviral particle happen to come in contact with the surface of a potential host cell. This is the point at which the envelope spikes play an essential role. These proteins have evolved to lock on to particular proteins that reside on the surface of host cells. In so doing, the viruses hijack the normal functions of these cellular receptors, which are to bring essential components into the cell or to receive messages from other cells in the form of diverse molecules. Some receptors are common to all cells, whereas others are limited to only a few cell types. The envelope proteins of

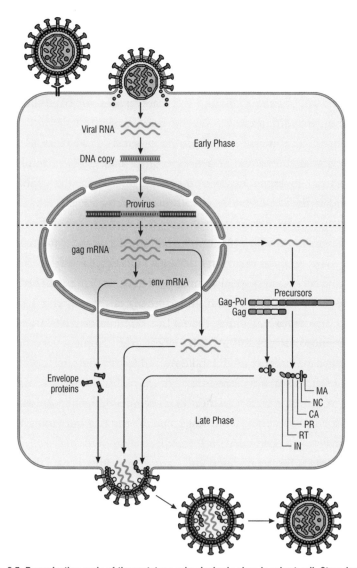

Viral RNA

Early Phase

DNA copy

Provirus

gag mRNA

env mRNA

Precursors

Gag-Pol
Gag

Envelope
proteins

MA
NC
CA
PR
RT
IN

Late Phase

Fig. 2.5 Reproduction cycle of the prototype avian leukosis virus in a host cell. Steps in the early phase are as described in Figure 2.3. In the late phase, RNA synthesized from the provirus serves both as new viral genomes and as mRNAs for production of the viral protein components as follows. About one-third of the viral RNA is spliced to form mRNA (labeled *env* mRNA) for the spikes that will be embedded in the viral envelopes. Two copies of the remaining full length RNA are incorporated in to each progeny particle. The rest of this RNA (labeled *gag* mRNA) directs synthesis of the Gag polyprotein precursor, which will provide the majority of core components for the newly formed virus particles (MA, CA, and NC). This mRNA also directs synthesis of a smaller of amount an extended, Gag-Pol polyprotein, which includes viral enzyme precursors (PR, RT, and IN). Following budding of the precursors from the cell surface, activation of the viral protease (PR) results in release of the structural proteins and enzymes and formation of mature, infectious particles.

each group of retroviruses have distinct receptor specificities that determine which cells they can enter. For example, the major receptor for the AIDS virus, the human immunodeficiency virus (HIV), is a cellular protein called CD4, found only on particular cells in our immune system.

Although there are variations on the theme for different retroviruses, and many of the details have yet to be worked out, by the 1990s, a general outline of their reproductive cycle in a host cell had been delineated. Two phases were distinguished in this cycle. The early phase includes all of the steps from attachment of a particle to the cell receptor to integration of retroviral DNA into the DNA of the infected cell. Production of RNA from the integrated retroviral DNA and subsequent assembly of new infectious particles comprises the late phase.

The early phase begins when a retrovirus particle is attached to a host cell and a component of the spike causes the viral envelope and outer cell membranes to fuse, allowing the capsid to be deposited into the cytoplasm. Reverse transcription begins shortly thereafter, using deoxynucleotide substrates available in the host cell to make retroviral DNA. After a complete DNA copy is produced, the retroviral integrase binds to the LTR ends, and this complex gains access to the host genome either by entering the nucleus through pores in the nuclear membrane or simply by waiting until the nuclear envelope breaks down and then reforms, during cell division. Once in the nucleus, the integrase binds to host DNA, in some known cases assisted by cellular proteins that normally associate with host DNA. The integration reaction is fairly promiscuous; while different viruses have distinct preferences, many sites in the host DNA can be targets for retroviral DNA integration. From this point on, the integrated viral DNA, Temin's provirus, masquerades as a cellular gene, coopting the cell's machinery for production of progeny virus particles or, as will be described later, transforming or affecting the host cell in a variety of ways.

During the late phase in the retrovirus reproductive cycle, synthesis of RNA from the integrated provirus is directed by signals within the LTRs, which recruit the required host cell transcription machinery and determine where synthesis of the viral RNA starts and stops. Sequences in the newly made RNA product, called splicing signals, then direct the host cell machinery to remove certain segments in about a third of these RNA molecules to produce separate mRNAs for the spike protein. Both spliced and unspliced viral RNAs are then transported to the cytoplasm. Two copies of the unspliced RNA will be encapsidated into each nascent virus particle as genomes. The remaining unspliced and spliced RNAs serve as messengers that are translated by the cellular ribosomes into viral polyproteins called Gag and Gag-Pol and Env, respectively. The Env proteins travel to the surface membrane of the host cell into which they are inserted, eventually to form "spikes" in progeny particles that will enable infection of a new cell. The Gag and Gag-Pol polyprotein precursors, as well as two RNA genomes, come together at the underside of the cell's surface to self-assemble into new particles that are shed via budding from the cellular membrane from which the retroviral envelope is acquired. After the nascent particles leave the cell surface, the viral protease snips the viral polyproteins, releasing the individual functional components. This final step in retrovirus "maturation" is marked by a morphological change in which the capsid is seen to condense.

As research progressed through to the 1990s, it became clear that the retroviruses are among the most widely distributed family of infectious agents. All animal species, including humans, are hosts for

Table 2.1 The Retrovirus Family

Genus	Examples[a] (Genome Type)[b]
Alpharetrovirus	Avian leucosis virus (simple)
Betaretrovirus	Mouse mammary tumor virus (complex)
Gammaretrovirus	Murine leukemia virus (simple)
Deltaretrovirus	Human T-cell lymphotropic virus (complex)
Episilonretrovirus	Walleye dermal sarcoma virus (complex)
Lentivirus	Human immunodeficiency virus (complex)
Spumavirus	Chimpanzee foamy virus (complex)

[a] Each genus includes numerous additional members and, in many cases, viruses that infect additional animal species.

[b] Simple genomes contain only the three conserved retroviral genes, *gag, pol,* and *env* (Figure 2.4); complex genomes include additional protein-coding sequences, often either in front or after *env.*

retroviruses. Although initially retroviruses were classified according to the similarities in their structure as seen in the electron microscope, each group has now been assigned to a specific genus based on nucleotide sequence and shared features of genome organization. The cancer-causing "oncogenic" retroviruses comprise five groups, depending on shared details of their genetic maps. Each group has been assigned to a distinct genus, named from the Greek alphabet: *Alpharetrovirus* through *Epsilonretrovirus*. Retroviruses that cause diseases of the central nervous and immune systems comprise a distinct genus known as the *Lentiviruses*. This genus includes the first retrovirus to be discovered, the equine infectious anemia virus, as well as one of the latest, the human AIDS virus, HIV. The third group of retroviruses is somewhat of a curiosity. They are called foamy viruses (genus *Spumavirus*) because of the bubbly appearance of infected cells. These retroviruses have an unconventional genome and lifestyle and, as yet, no known pathology.

Study of the pathogenic retroviruses, as well as knowledge of the unique behavior of these agents as parasites of their host's genome, opened new avenues for understanding the causes of cancer and other diseases in both animals and humans, as will be described in later chapters. Moreover, because DNA copies of any cellular or viral RNA can be made with reverse transcriptase, the isolated enzyme became an indispensable tool in genetics and the burgeoning field of biotechnology. Such knowledge and unique technical capabilities also brought new questions into focus. Where did the retroviruses come from? And how unique really is reverse transcription in the biological world?

The Origin of Retroviruses

Knowledge of the relative simplicity of viruses, but diversity of their genome composition and structure, led to competing theories about the origin of these agents. Some scientists speculate that viruses were formed from bits of DNA or RNA that "escaped" from living cells. Others believe that viruses evolved independently and, perhaps, even prior to the first single-cell organisms on our planet. To date, no clear explanation for the origin of all viruses exists. However, as will be described below, most scientists believe that the very first genes comprised RNA. Therefore, probing the origin of viruses with RNA genomes, especially the retroviruses, has offered unique insight into primordial life on Earth.

In the Beginning . . .

The question of how life began on Earth has mystified humans throughout the ages. Numerous myths and diverse religions offered answers not only for who was responsible but also how life transpired. The question of "who" unfortunately is not yet accessible to analysis by scientific methods. But scientists have attempted to address the "how" ever since the essential roles of nucleic acids and proteins in biological systems were identified.

Geologists estimate that our planet is about 4.5 billion years old and life arose 3.8 billion years ago, at a time when Earth was a most inhospitable place. The planet's surface was bubbling with volcanic eruptions and pummeled with acid rain, while the atmosphere contained little or no oxygen and no ozone layer for protection against radiation. Modern practitioners of paleontology, the study of ancient life, have used a variety of methods, including biochemistry and geochemical methods, to determine that simple single-cell organisms, like bacteria, were the first living forms on Earth. However, some evolutionary biologists argue that primitive virus-like replicating forms may have pre-ceded these entities. The simplest multicellular organisms with which we are familiar only appeared in the past 570 million years. That leaves more than 3 billion years for creatures to evolve from simple cells and for viruses to contribute to such evolution. As we can't go back in time, it is difficult to prove or rule out any present hypothesis about how life got started on our planet. That limitation aside, evidence for the fea-sibility of certain critical primordial reactions has been acquired over the years in which RNA and, perforce, reverse transcription figure prominently.

Most scientific theories start with the premise that life on Earth began in some sort of primitive chemical soup containing the minimal essential ingredients. The forces that might have brought such ingredi-ents together to form molecules like those in living cells are un-known, but early conditions on Earth in tidal pools, on the surface of clay sediments, or in deep-sea hydrothermal vents are thought to have been able to foster such reactions. Stanley Miller and Harold Urey at the University of Chicago were the first investigators to obtain support for the idea that some of the basic building blocks of life could have been formed from the rudimentary ingredients present on early Earth. In their famous experiments of 1953, simple gases thought to be present at that time in the atmosphere, methane (CH_4), ammonia (NH_3), and hydrogen (H_2), were enclosed in an apparatus that included a series of connected flasks. One flask containing water (H_2O), repre-senting the ocean, was heated to add water vapor to the reaction. Electrical sparks were then fired continuously through the gases to simulate lightning as a source of energy.

After a week of operation, 10 to 15 percent of the carbon enclosed in the Miller / Urey apparatus had been incorporated into organic com-pounds, including thirteen of the twenty major amino acids that make up the proteins in living organisms. This amazing result inspired nu-merous similar experiments in what has been called chemical evolution. Later studies showed that the nucleotide bases of RNA (uracil) and DNA (adenine) could also be produced under conditions assumed to simulate early Earth's environment. These fascinating explorations in prebiotic science are not, however, without critics. Because the exact

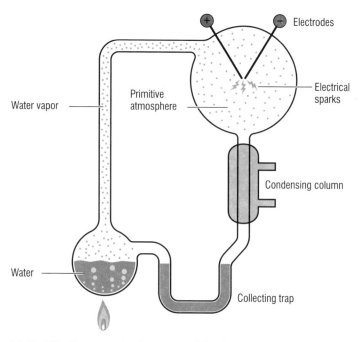

Fig. 3.1 The Miller-Urey apparatus. Components of the pioneering experiment in chemical evolution are identified.

conditions on primordial Earth are not now (and may never) be known, interpretations of the results can certainly be questioned. Moreover, some scientists have proposed that the ingredients for initiation of life, or even some form of life itself, might have been delivered to Earth in bombarding comets or other objects from outer space. In support of this idea, ribose, the sugar component of RNA, is formed when frozen mixtures of water, methanol, and ammonia are subjected to UV irradiation, conditions believed to simulate effects on the interstellar ice present on such celestial objects.

The RNA Origin of Life Hypothesis

Two essential qualities that distinguish living entities from inanimate objects are the ability to reproduce and adapt to the environment (evolve). There are several reasonable scenarios for how these two properties might have arisen from ingredients such as the amino acids and nucleic acid bases formed by chemical reactions on early Earth, but there is no current consensus. There *is* general agreement, however, that RNA played a major role in an early stage of the evolution of life, as we now know it. In this "RNA world," as it is called, primitive RNA chains are thought to have arisen spontaneously, with support and energy from some promoting feature of the early Earth environment. Such chains could then multiply by serving as templates for base pairing with available nucleic acid substrates. Successive rounds of this mode of replication would be the basis of genetic continuity for RNA chains. One can think of such chains as "selfish" mini-genes whose main function is their own reproduction. Occasional mistakes in base pairing (mutations) would then be responsible for driving their evolution. The most stable and the most efficient at making copies would gradually outnumber all the rest. It is also conceivable that small molecules, such as peptides, or various metals and other inorganic substances in the environment may have supported such reactions. But in this world, RNA served as both repository of information (genes) and catalyst (enzymes).

The general notion that RNA played a central role in life's origin was first publicly stated in a 1962 presentation by Alexander Rich: "We postulate that the primitive polynucleotide chains are able to act as a template or as a somewhat inefficient catalyst for promoting the polymerization of the complementary nucleotide residues ... in a primitive environment in the absence of protein catalysts."[1]

Rich's ideas did not gain much traction perhaps, as he pointed out, because there was a lack of relevant information. But similar speculations arose from others. By the early 1960s, RNA molecules were known to function as messengers from genes encoded in DNA (messenger RNA [mRNA]), as amino acid–carrying information adapters (transfer RNAs [tRNAs]), and as critical components of the ribosomes (ribosomal RNAs) where amino acids are linked together in a specific sequence to make proteins. The idea that life developed from nucleic acids was proposed independently in 1967 by an American, Carl Woese, and in 1968 by British scientists Francis Crick and Leslie Orgel.

Carl Woese was trained as a biophysicist. By the mid-1960s, he had become fascinated with microbes and the process of evolution. Woese conceived the notion that evolutionary relationships among members in the tree of life might be determined by studying shared, essential genes. Protein synthesis is critical to all cellular life, and a consideration of available evidence suggested to him that the most primitive ribosomes were likely to be made up entirely of RNA. Woese reasoned that changes in the sequences encoding ribosomal RNAs would occur at a measurable rate over time, as genes were duplicated during cell division. By comparing these sequences among organisms, it would be possible to identify evolutionary relationships and determine how much time elapsed in the appearance of distinct species, a kind of molecular chronometer. Unfortunately, tools for analyzing RNA were rudimentary at the time, and it would be another decade until the reclusive Woese and his postdoctoral associate George Fox could gather enough information to convince themselves of the validity of their approach. In the course of their studies, they discovered that certain microbes, assumed to be bacteria, contained ribosomal RNA that looked nothing like that of bacteria. This amazing discovery was reported in a landmark paper of 1977, in which Woese argued that these microbes are members of an entirely new branch in the tree of life, which he called the Archaea (Latin for "primitive").[2] Woese's ideas were quite radical and, as previously noted, with most exceptional discoveries, many in the biological world were skeptical about his work. His reclusive nature and bitter reaction to this skepticism earned him a reputation as a crank. While it took another decade for the community to recognize the importance of his work, construction of phylogenetic trees based on gene sequences is standard today. Furthermore, in recognition of their diagnostic importance, the first genes to be sequenced in any new microbial species are those that encode ribosomal RNA. Woese is now known as the scientist who rewrote the tree of life, as well as one of the most important, early proponents of the RNA world hypothesis.

Other early notable proponents of an RNA world were the RNA Tie Club members Francis Crick and Leslie Orgel. These two scientists independently proposed in 1968 that nucleic acids could act as catalysis and that RNA might, therefore, promote its own replication. Like Woese, they also speculated that the first ribosomes may have been made entirely of RNA, but neither searched for proof in contemporary organisms, believing that such primordial reactions were unlikely to be in evidence today. In the following decades, several lines of research reinforced the concept of a primordial RNA world. For example, the RNA component of contemporary ribosomes is responsible for setting up almost all of the necessary contacts between mRNA and tRNAs during protein synthesis. Furthermore, the site for joining amino acids to form proteins lies deep within the central core of ribosomal RNA. As the numerous protein components of ribosomes are bound primarily to the outside of this protein-synthesizing machine, it is believed they were probably acquired over the course of evolution to maintain a proper architecture for the RNA. By using test-tube methods to simulate nucleic acid evolution, it has been possible to select RNA molecules performing a wide range of different catalytic reactions.

Discovery of Self-Cleaving RNA Perhaps the most compelling support for the existence of an RNA world has been the discovery of naturally occurring, self-splicing RNAs called "ribozymes." In these RNAs, the ability to cleave RNA chains and rejoin the broken ends is intrinsic to their structure and independent of any protein. Thomas Cech and colleagues discovered the first ribozyme in 1982, in a tiny pond-dwelling, single-cell organism called *Tetrahymena*. Cech wanted to study the structure and expression of the gene for ribosomal RNA, for which this organism seemed most advantageous. *Tetrahymena* cells are somewhat unique in having not one but two nuclei. One, called the micronucleus, contains the five chromosomes passed on to progeny. The second, a large macronucleus, is formed at a particular point in the *Tetrahymena* life cycle by fragmentation of the genome to produce 200 to 300 minichromosomes that then undergo massive duplication. Most important for Cech, the minichromosome encoding ribosomal DNA exists in approximately 10,000 copies in the *Tetrahymena* macronucleus.

Cech's initial analysis showed that the *Tetrahymena* ribosomal gene included a short section not present in the final ribosomal RNA product and must be removed from its precursor RNA. Thinking that this simple system might be ideal for identifying a protein that removed this intervening sequence (abbreviated IVS), he had one of his students purify the ribosomal precursor RNA and look for such a protein in a cellular extract. When the student reported the IVS was removed in both the presence *and the absence* of added cellular extract, Cech thought that there must be some mistake. However, five careful repeats of the experiment yielded the same puzzling result. It appeared that excision of the IVS from the precursor RNA did not require a protein. Perplexed, it took two years and numerous experiments for Cech and his students to convince themselves that the splicing reaction they were studying was not produced by a minor protein contaminant in the precursor RNA but was instead intrinsic to the RNA itself. Their biochemical analysis of the tiny IVS RNA produced another surprising result: during excision from the precursor, it acquired an extra guanosine nucleotide (G) residue at one of the cleaved ends. The mechanism they ultimately deduced for the self-splicing reaction of *Tetrahymena* included formation of a twisted structure, which brought two strands of the ribosomal precursor RNA into close proximity to assemble an "active center." The IVS is then spliced out of the precursor with the assistance of a free G nucleotide, and the remaining ends of the precursor RNA are joined. In a final reaction, a G residue at the other end of the *Tetrahymena* IVS attacks an internal location near the end with the new G nucleotide, forming a small RNA circle. Although self-splicing is a relatively rare phenomenon, additional examples have since been found in RNAs of bacteria, plants, and animals. On the other hand, as described below, another kind of RNA splicing is very common and, indeed, is essential for formation of most mRNAs in higher organisms.

Split Genes and Spliceosomes It was a great surprise in the late 1970s when scientists discovered that genes in some eukaryotic DNA viruses are not arranged in a contiguous linear segment, as they are in bacteria. This Nobel Prize–winning discovery was made by two groups studying the DNA and mRNA of human adenoviruses: one led by Philip Sharp at MIT, with Susan Berget as a major collaborator, and a second led by

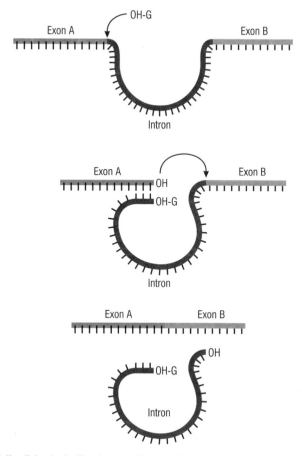

Fig. 3.2 Self-splicing by the *Tetrahymena* ribozyme. Two exons in the RNA chain are shown in a straight line with the intron looped out below. Following base pairing of sequences in the intron with those at the end of exon A, the first cleavage reaction is facilitated by a molecule of guanosine triphosphate (G). A second cleavage reaction results in joining of exons A and B with release of the intron.

Richard Roberts, working with Loise Chow and Tom Broker at Cold Spring Harbor. The two groups discovered independently that the mRNAs produced by these viruses included sequences encoded in different parts of the viral genome. The mRNAs were made by splicing together noncontiguous segments from longer precursor RNAs. Shortly thereafter, development of DNA cloning and sequencing methods revealed that most genes in eukaryotic organisms contain alternating protein-coding and noncoding sequences, signifying that the RNAs from these genes must also be spliced. A specialized machine in the cells' nuclei, a "spliceosome," was shown to perform this processing. Spliceosomes, made up of both proteins and RNA molecules, cut out the noncoding RNA sequences (called "introns") in precursor RNAs and join the protein-coding sections ("exons") to form mRNAs. However, as with the ribosome, it is the RNA components of the spliceosome that catalyze the essential chemical reactions.

Other RNA-Protein Collaborations A different type of protein-RNA association that employs a catalytic RNA was described in 1983, just a year after report of the *Tetrahymena* ribozyme from Cech's laboratory. Sidney Altman and his colleagues at Yale University had been studying a bacterial enzyme called RNase P, which removes extra sequences from precursor tRNA molecules and other cellular RNAs. In 1978, Altman reported that this enzyme consists of one protein and an RNA molecule that played some essential role in the reaction. Altman noted at the time that he and his colleagues "did not have the temerity to suggest, nor did we suspect, that the RNA component alone of RNase P could be responsible for its catalytic activity. The fact that an enzyme had an essential RNA subunit, in itself, seemed heretical enough."[3] However, one of his students discovered the RNA alone *could* carry out the reaction catalyzed by RNase P, under appropriate conditions. It is now accepted that the protein component in this enzyme functions as a supporting scaffold and not a catalyst. The RNase P enzyme in humans includes an RNA molecule similar to that in the bacterial enzyme, but the human enzyme contains more than one protein. In 1989, Altman and Cech were awarded the Nobel Prize in Chemistry "for their discovery of catalytic properties of RNA."

Additional examples of RNA working together with proteins have since been found in nature. Such assemblies are called ribonucleoproteins, or simply RNPs. The first RNPs to be identified were the complexes in which mRNAs are processed or transported to the cytoplasm for ribosomal translation. Like ribosomes and spliceosomes, such RNPs are now presumed to be present-day remnants of an ancient transition

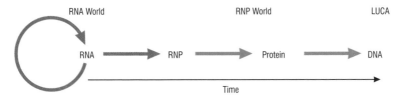

Fig. 3.3 Successive steps in the transition from an RNA world to a DNA world. LUCA stands for the last universal common ancestor, a single cell with a DNA genome that arose 3 to 4 billion years ago and from which all life on Earth has evolved.

from the RNA world to one in which proteins were recruited by RNAs to enhance function and efficiency.

Despite the ancestral role of RNA as the very first carrier of heredity, in today's biosphere, RNA genomes are only found in viruses. For all other life forms, *the* major vehicle for storage of genetic information is DNA. The transition to DNA as a universal genetic reservoir is generally perceived as having occurred in several steps, first from an RNA world to an RNP world and eventually to the present world based on DNA.

Transition to the DNA World

Why use DNA, if RNA could do the job? What factors could have driven this transition during evolution? One likely driver is the fact that DNA is more stable than RNA. Even though the components in these two nucleic acid polymers are very similar, changing the sugar in a nucleic acid chain from ribose (RNA) to deoxyribose (DNA) is estimated to increase its stability about 100-fold. Because one break in an RNA chain can destroy the continuity of its genetic content, there is therefore a greater limit to the amount of genetic information that can be stably maintained in RNA versus DNA. The longer the RNA chain, the higher the probability that such a break will occur. In addition, a single-strand break in double-stranded DNA need not be a catastrophe, because the complementary strand may be used as a template for its repair—offering another source of genome stability.

There are large gaps in our understanding of exactly *how* or *when* the evolutionary transition to a DNA world might have occurred. Some speculate that the changeover started even before cells or viruses existed. However, the discovery of the retroviral reverse transcriptase provided our first glimpse of how such transitions might have been facilitated. Genetic information initially present in a single RNA chain can indeed be copied efficiently into double-stranded DNA by this enzyme. Although previously thought to be unique to retroviruses, reverse transcriptases were later found in other viruses, including the human hepatitis B virus and in a family of viruses that infect plants. Moreover, it is now clear from studying the genomes of bacteria and Archaea that reverse-transcribing enzymes arose very early in evolution, as did enzymes that can insert one DNA segment into another, like the retroviral integrase. The DNA of organisms in all domains in the tree of life is inhabited by sequences called transposable genetic elements, which use such enzymes to insert themselves or their DNA replicas into their "host" genomes. These elements comprise a major component of most genomes, including our own. Called "jumping genes,"

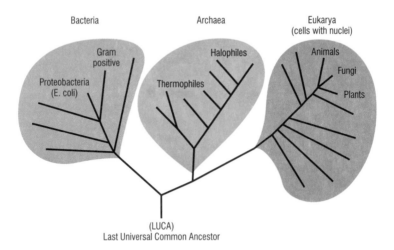

Fig. 3.4 The tree of life. Bacteria and Archaea comprise the prokaryotes, single-celled organisms that lack defined nuclei. The cells of all eukaryotic organisms have defined, genome-sequestering nuclei.

and first thought of as "junk" DNA, they are now known to have profound effects on evolution.

Transposable Genetic Elements

One cannot discuss jumping genes without noting the extraordinary
work of Barbara McClintock. While still a graduate student in the Agricultural School at Cornell University, McClintock developed refined
techniques to identify each of the ten chromosomes in the cells of corn
plants. She gained recognition early in her career as an exceptional
cytogeneticist for studies on the structural relationship of these chromosomes and inheritance of genetic traits (Chapter 1, Table 1.1). A tiny,
yet fiercely independent and private person, she found a position most
congenial to her solitary style of research at the Cold Spring Harbor
Laboratory in the early 1940s. The laboratory is in a bucolic setting on
the north shore of Long Island about 40 miles from New York City and
is surrounded by hills and lovely mansions. McClintock was supported
by the Carnegie Institution of Washington, in the Department of Genetics, headed by Alfred Hershey. Free from teaching or the administrative duties typical at a university, she grew experimental corn plants
on the laboratory grounds during the summer. She was also able to ab-
sorb and enjoy the intellectual stimulation available at the laboratory,
a kind of "summer Mecca" for the world's most eminent geneticists. In
winter, alone and with little distraction, she analyzed the effects of her
crosses on chromosome structure in the microscope. It was while ex-
amining chromosomes in the late 1940s that McClintock happened
upon a totally unknown and completely unexpected phenomenon.

During investigations of the genetic basis for different color patterns
in kernels of maize, commonly known as Indian corn, McClintock discovered that color variations were associated with visible breaks at a
specific site on maize chromosome number 9. Additionally, the appearance of such breaks depended on a particular locus at the other end of
the same chromosome, which she surmised must contain a "controlling

element" because it affected the expression of the gene that she was studying. She ran into trouble, however, when trying to map the element, as she found that it actually *changed positions* in the chromosomes of progeny from her crosses. This was a truly jarring discovery. Even though the chemical nature of genes was still uncertain, genomes were assumed to be completely stable features of cells, unchanging from one generation to another. The idea that bits of genetic information could move around was simply unimagined at that time. McClintock was fully aware that her findings challenged the current ideas about inheritance and were contrary to common conceptions. However, she could imagine how the controlling element would produce the variations in color patterns that she observed and could also grasp the broader implications of its transposition to different locations. Such movement could be responsible for producing new mutations, or rearrangements that affect the evolution of genomes, or produce genetic changes that control steps in the development of organisms.

McClintock documented her initial results in 1948 in the *Annual Report of the Carnegie Institution,* which was a collection of selected highlights of work done in the past year by each laboratory. A formal report of her experiments was published in the *Proceedings of the National Academy of Sciences of the United States of America* journal in 1950. Although not disposed to self-promotion, she now felt compelled to present her remarkable findings and insights publicly to the scientific world. The forum she chose was the Cold Spring Harbor Symposium, a premier meeting for biologists and geneticists. Unfortunately, the results fell far short of her hopes. While clearly of genius caliber, as proven with her election to the National Academy of Sciences in 1944 and leadership as president of the Genetics Society of America in 1945, McClintock was not a practiced "communicator." She had an idiosyncratic, nonlinear way of speaking, with a tendency to interrupt herself with edits and footnotes, as she spoke. Moreover, not only was the subject matter dense and difficult to follow, but the complexities of corn genetics also were foreign to most people in the audience who were focused on the simpler genetics of bacteria or bacteriophage systems. The few who could follow wondered if the properties she described were some anomaly of her system. There seemed no way at the time to integrate her results into a coherent scheme for inheritance. No questions or comments followed her talk, only total silence. One can only imagine the colossal letdown for McClintock. Although frustrated at her failure to convince her peers of the significance and broader implications of her observations, she had no doubts. McClintock continued her solitary studies into the next decade, discovering additional examples of transposable genetic elements in the chromosomes of maize. She tried twice more to present her work to an audience of colleagues in 1956, at meetings held in the nearby Brookhaven Laboratory and at Cold Spring Harbor, with no more satisfying results. Thinking further attempts to be fruitless, she stopped publishing her results in all but the Carnegie Institution's *Annual Reports.* Nevertheless, she remained confident that the implications of her work would eventually become apparent, when similar observations would be made in other systems.

As McClintock predicted, beginning in the late 1960s and continuing into the end of the century, a new appreciation of the dynamic properties of DNA arose from numerous directions. The integration of DNA of temperate phages into the genome of the bacterium *E. coli,* which establishes the lysogenic state, was now seen as one example of genetic transposition. The DNA of another *E. coli* bacteriophage, called mutator (Mu), was found to be integrated at random sites in its host genome where it could induce mutations by interrupting host genes. Mu was therefore especially reminiscent of McClintock's transposable elements and confirmed many of her predictions. The application of new methods allowing genes to be isolated and manipulated, along with advanced nucleic acid sequencing technologies, subsequently revealed the presence of transposable genetic elements in many other microorganisms and animals. Although still working alone, McClintock kept track of these advances, discussing their implications with those in the field who came to visit or who worked at Cold Spring Harbor. As examples of "jumping genes" began to turn up virtually everywhere, McClintock's experiments eventually came to be considered "classics," experiencing a renaissance of sorts in the next two decades. She came to be revered by many of the young leaders in the fields of molecular biology and genetics. Like Rous, Barbara McClintock was officially recognized as a scientific forebearer toward the end of her lifetime. At age eighty-one, she received the 1983 Nobel Prize in Physiology or Medicine "for her discovery of mobile genetic elements."

There is now widespread appreciation of the dynamic features of genomes, and "mobile DNA" has become a popular and highly populated field of study. We have come to appreciate that some genomic changes are actually programmed to occur in certain cells or under certain conditions. These reactions can involve very small or quite large segments of DNA. The rearrangements and recombination reactions that generate the vast variety of antibody genes in our blood cells is one well-studied example of such programming; the gene shuffling that produces two different types of yeast cell that can then mate and exchange sequences with each other is another. Other dynamic alterations are promoted by a large variety of transposable elements that can move themselves within and between genomes. Such autonomous elements fall into distinct categories: transposons that jump as DNA elements and those that jump via RNA intermediates, which are copied into DNA before or while moving to a new location, the retrotransposons.

Transposons Are Us

When the first human genomes were sequenced in 2001, it was an enormous shock to discover that the protein-coding exon sequences containing the information for all of the proteins in our bodies account for only about 1 percent of our DNA. In mind-boggling contrast, more than half of our DNA is made up of transposable elements or repetitive sequences derived from them. DNA transposons comprise approximately 3 percent of human DNA, while a whopping 42+ percent of our DNA consists of retrotransposons. Recent computational analyses indicate that approximately half of the remaining portion of our genome, considered uncharted "dark matter," is also composed of previously unrecognized repetitive sequences derived from transposable elements. While there are still many mysteries to be explored in the

Table 3.1 Major Classes of Transposable Elements in the Human Genome

Type	Number of Copies	Length (Base Pairs)	Percentage
DNA transposons	300,000	80–3,000	3
RNA transposons			
Non-LTR			
LINEs	860,000	6,000–8,000	21
SINEs	1,500,000	100–300	13
Pseudogenes	1–100	variable	<0.5
LTR (HERVs)	450,000	1,500–11,000	8

Source: E. S. Lander, et al., "Initial Sequencing and Analysis of the Human Genome," *Nature* 409 (2001): 860.
Note: LTR = long terminal repeat; LINE = long interspersed element; SINE = short interspersed element; HERVs = human endogenous retroviruses.

fantastic landscape of our DNA, it is hard to escape the rather humbling conclusion that most of our genome is nothing more than a collection of transposable elements and their by-products, which have accumulated over eons of evolution. And we are not alone; an astounding 50 to 90 percent of the genomes of many animals and plants are made up of such elements.

DNA Transposons

DNA transposons are found in all branches in the tree of life, but they are most common in bacteria. Some are moved simply by being excised from one site in a genome and inserted into another, a mechanism called "*cut* and paste." Others are like rubber stamps: replicas produced by DNA-directed DNA synthesis are inserted into various additional locations. These movements result in the net increase of what are known as "*copy* and paste" transposons. Jumping by both types of DNA transposons is made possible by proteins that are encoded in the elements (enzymes called transposases or integrases), which typically bind to short DNA sequences repeated at either end of the element. To note, years after McClintock discovered the first jumping gene in corn chromosomes, the work of others revealed it to comprise a "cut-and-paste" DNA transposon that encodes its own transposase enzyme.

Some DNA transposons are able to pick up and mobilize genes that encode other functions, including cellular proteins that can block the action of antibiotics. The transfer of such elements from one bacterial strain to another, usually on rings of DNA called plasmids, is responsible for much of the rapid appearance of multidrug resistance in pathogenic bacteria. Acquisition of such resistance is of obvious benefit to the pathogen; it can now flourish in a sea of drugs, but this is clearly a disaster for an infected human host—and an ever-increasing problem for the medical community.

Evolution has also taken advantage of DNA transposons for purposes that have turned out to be useful for human beings. For example, we now know that the enzymes in our blood cells that mediate the programmed rearrangements leading to antibody production, as well as the signal sequences that guide these processes, were coopted from an ancient DNA transposon(s) that inhabited the genome of vertebrate species some 500 million years ago. In this respect, we owe our ability to fight off infections to jumping genes.

Retrotransposons

Like some DNA transposons, the retrotransposons are rubber-stamped into various locations, thereby increasing their number, but in this case replicas are made by reverse transcription of their RNAs. The retrotransposons of bacteria are presumed to be among the most ancient jumping genes. The best studied among them, the group II "retrointrons," produce mRNAs that are both ribozymes and encode a reverse transcriptase. After the reverse transcriptase is produced, the retrointron segment cuts itself out of the mRNA and binds to the reverse transcriptase. This complex then attaches to a new site in the bacterial genome where the retrointron RNA is copied into DNA by the reverse transcriptase. These bacterial retrointrons are presumed to be the ancestral progenitors of the introns that are recognized and removed by spliceosomes described previously and also progenitors of the most common retrotransposons in humans and other mammals.

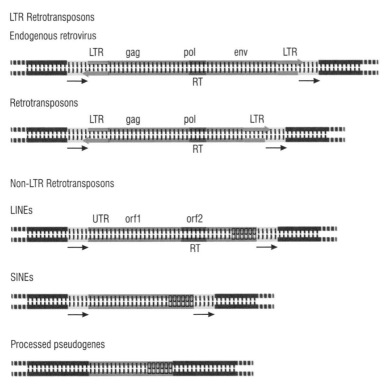

Fig. 3.5 The retrotransposons. The maps show the integrated DNA forms. The LTR retrotransposons are characterized by flanking direct repeats of host DNA (depicted in yellow) formed by duplication of a host target sequence during integration and long terminal repeat sequences (LTRs; in red), which include signals for the initiation and termination of RNA synthesis. The non-LTR retrotransposons all contain long stretches of A-T base pairs at one end. SINEs, and the autonomously transposing LINEs, are also flanked by repeats of host DNA target sequences.

As noted above, retrotransposons comprise a major portion of our own DNA and that of higher organisms. Most of the retrotransposons that inhabit our genome are stationary and inactive today. They are essentially the fossil remnants of events that occurred many millions of years ago in the DNA of ancestral species. The functional features of these elements have been eroded away by numerous random mutations and deletions over time; in some cases, only bits of identifiable sequences remain. However, as will be described below, some elements are still active today or show evidence of having been active in the recent past.

Two major classes of retrotransposons are distinguished by their distinct structures and jumping mechanisms: those that contain long terminal repeats (LTRs) (like the retroviruses) and those without LTRs.

LTR Retrotransposons This class of retroelements includes sequences derived from retroviral DNAs that were inserted into germline cells (eggs or sperm) following infection of ancestral species millions of years ago. Once in the germline, they were passed on through subsequent generations. These elements are called endogenous retroviruses, abbreviated as HERVs in humans. Full-length endogenous retroviruses contain recognizable *gag, pol,* and *env* retroviral genes (as illustrated in Figure 2.5). However, numerous isolated fragments of HERV genes, or their solo LTRs, are also found in our DNA. With the possible exception of one, called HERV-K, the human elements are inactive, having accumulated numerous deleterious mutations during evolution. It is suspected that HERV-K might have been active in the recent past, however, because individuals have been found who lack HERV-K sequences in particular locations. As will be discussed in Chapter 4, endogenous retroviruses have had a profound impact on evolution, including that of humans.

Some LTR retrotransposons are not derived from viruses. These elements encode the equivalents of the retroviral *gag* and *pol* genes, and they make virus-like particles in the cytoplasm of the cells that they inhabit. However they have no *env* gene that would allow them to exit their host cell and infect other cells. These elements can only jump around the genome of the cell that they inhabit. It is generally believed that retroviruses are derived from ancient LTR retrotransposons that became infectious after acquisition of an *env* gene and related mechanisms for budding from cells. The LTR retrotransposons and retroviruses appear to be the most highly evolved retrotransposable elements. Their reverse transcriptases are related to that of their cousins, the LINE elements described below, but their LTRs and structural genes were accumulated from diverse sources during their evolution, which is likely to have begun more than a half billion years ago—and continues to this day.

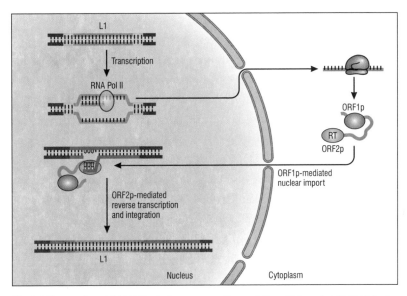

Fig. 3.6 Propagation of L1 LINE retrotransposons. Transcription of integrated L1 DNA by the host cell RNA polymerase produces LI RNA molecules (depicted in green). L1 RNA is transported to the cytoplasm and translated by ribosomes to produce the two L1 proteins (ORF1p and ORF2p). The L1 RNA binds these proteins and the complex is transported back into the nucleus where the polyA tails at the end of the RNA can form base pairs with Ts in the DNA at a new target site in the host DNA (represented by a light brown segment). The L1 RNA is then reverse transcribed by ORF2p, and the newly formed L1 DNA is integrated into the host DNA. The L1 DNA is flanked by direct repeats formed by duplication of the new target sequence (depicted in orange) during integration.

Non-LTR Retrotransposons The major class of non-LTR retrotransposons consists of elements called LINEs (for long interspersed elements). Like the ribozyme-encoding bacterial retrointrons, LINEs insert a DNA copy of their own mRNA into various sites in genomic DNA using their own reverse transcriptase enzyme. The LINE-1 (L1) elements in the human genome are members of a large family that have been jumping and evolving in mammals for more than 100 million years. The human genome contains about 500,000 copies of L1, but only an estimated 80 to 100 are active in any one individual. These are the only autonomous transposable elements known to be active in humans. However, these elements have retained the ability to mobilize not only their own mRNA but also the RNAs made by other repetitive elements that lack this ability, called SINEs (for short interspersed elements). Evidence for mobilization by LINEs has also been discovered in sequences in the human genome called "pseudogenes" (i.e., fake genes that do not encode proteins), which are nonfunctional DNA copies of mRNAs from protein-coding genes that have been inserted into distant locations in the genome. Copies of mRNAs from some viruses, other than retroviruses, have also been inserted into the genomes of our ancestral progenitors by ancient LINE elements. Discovery of these viral pseudogene fossils in human DNA provided the first estimates of the age of several viruses that are still circulating today. We now appreciate that they are the descendants of strains that were infecting our progenitors long before humans appeared on our planet.

Large-scale DNA sequencing studies have shown that the DNA of any two unrelated humans will differ by the presence or absence of L1 insertions at numerous sites. While L1 appears to be inactive in most normal adult tissues, it *is* active during early fetal development. The rate of L1-mediated jumping has been estimated at about one event in every 95 to 270 human births. Depending on the landing site, some of these jumps may be innocuous or even advantageous, providing new fodder for evolution. But we know that other jumps can have serious consequences, causing debilitating mutations in the genes that are invaded.

The first report of disease-related mutations caused by insertion of L1 DNA appeared almost thirty years ago during a screen of 240 unrelated males with hemophilia, the "bleeder" disease. In two individuals, independent L1 insertions were found to disrupt the gene for the essential blood-clotting protein, factor VIII. None of this critical protein was produced in either individual, and the disease was severe in both. Hemophilia is a sex-linked disease, normally passed on to male offspring though the maternal X chromosome. Because no abnormalities were found in the factor VIII gene of their parents, it was concluded that the debilitating L1 jumps must have occurred either in a parental germ cell or after conception of the affected individuals. Other human diseases, including various cancers and autoimmune disorders, have been associated with L1-mobilized transposition of itself or various SINEs into critical protein-coding genes. The deleterious insertion of a new pseudogene by L1 has also been detected. A total of 124 cases were documented by 2016, but this number is likely to increase, as whole-genome sequencing becomes a standard for medical diagnoses in the future.

The one normal tissue in which L1 is known to be unusually active is in the human hippocampus, the portion of the brain that controls memory, learning, and emotion. Much of this jumping appears to occur during early development of the human brain. While there is controversy concerning the actual rates (estimates vary from 0.1 to 80 new insertions per neuron), the genes most active in this region of the brain, and therefore most critical for cognition, appear to be preferred landing sites for L1 transpositions. Although data are still lacking, it is not hard to imagine that such events could have an impact on the functional properties of the brain, contributing perhaps to some of our personality quirks and distinctions or, in some unfortunate cases, to neurological abnormalities or disease.

And in the End(s) . . .

In the first half of the twentieth century, pioneering cytogeneticists, including Barbara McClintock, noticed that the natural ends of chromosomes, called telomeres, are somehow special. Whereas ends created by breaks anywhere along the linear chromosomes were quick to fuse to themselves or to other chromosomal fragments, telomeres never became attached to one another or to random breaks. Later, when the mechanism of DNA replication was solved, another mystery concerning chromosome ends arose. It was deduced from details of the mechanisms for DNA synthesis that only one strand of DNA could be copied to its very end by DNA polymerases. Consequently, it was estimated that 25 to 200 base pairs of DNA should be lost in each replication cycle and cell division, with a concomitant loss of heritable information. Yet this does not ordinarily happen. Both mysteries were solved during the 1980s through elucidation of the unique composition of telomeres and discovery of a cellular reverse transcriptase enzyme that was responsible for their formation.

As an example of convergence that happens not infrequently in science, the solution to the mystery of telomeres started with *Tetrahymena,* the very same organism chosen by Cech and for related reasons. A major player in this story, Elizabeth Blackburn, was born and educated in Australia and received her doctoral degree with Fred Sanger in the Laboratory of Molecular Biology, Cambridge, England, who was at the time pioneering methods for nucleic acid sequencing. She then joined the laboratory of Joe Gall at Yale University in 1975 as a postdoctoral fellow. Gall and colleagues had recently discovered the abundant, ribosomal gene-carrying minichromosomes in *Tetrahymena,* and Blackburn was anxious to apply the methods she had acquired in Sanger's lab to determine the composition of the telomeres in these minichromosomes. Using these and a variety of other techniques, she discovered the most unusual properties in these telomeres. They were not only heterogeneous in length but also composed of fifty to seventy repeats of the simple sequence CCCCAA. After reporting her amazing discoveries, Blackburn went on to set up her independent laboratory where she would continue her investigation of these unusual structures.

Jack Szostak had recently established his laboratory at Harvard Medical School to study the mechanisms of DNA recombination in yeast cells when he learned about the unusual composition of *Tetrahymena* telomeres at a conference in 1980 that he and Blackburn attended. Szostak was immediately intrigued by her findings and wondered if these telomeres could be replicated in yeast cells. Although neither really expected this experiment to work because yeast and *Tetrahymena* are so distantly related, the tools were available to test the idea and they decided to give it a try. Szostak stitched a DNA fragment from the end of one of the ribosomal RNA-containing *Tetrahymena* minichromosomes supplied by Blackburn to the end of a fragment of yeast DNA that he knew could replicate. He then introduced the chimeric construct into a yeast cell. To the amazement of both, this construct replicated as a linear chromosome in the yeast cells with telomeres at the ends. It was clear from this experiment that the *Tetrahymena* telomeres were recognized and replicated in yeast cells, although the repeated sequence produced at the ends of these chromosomes was the irregular G(1–3)T typical of yeast telomeres. These experiments were especially important because they established the universality of mechanisms for recognition and maintenance of telomeres. Later work by Blackburn in Berkeley and Szostak's lab in Boston showed that such maintenance is essential for cellular viability. When this function was blocked, telomeres became shorter with each cell division, and their ultimate disappearance was associated with cell death.

The inspiration of where to look for an enzymatic activity that produces telomeres came to Blackburn in a letter from Barbara McClintock. In response to a query from Blackburn, McClintock described some of her earlier unpublished experiments, which suggested that such an activity might exist in embryonic tissue of corn plants. Blackburn reasoned that the stage in which the minichromosomes of *Tetrahymena* were formed was likely to be an analogous point in this organism's development. Accordingly, that is where the attention of her graduate student, Carol Greider, was directed when she joined Blackburn's lab in 1984. After some trial and error, Greider developed an assay in which a short DNA fragment corresponding to the eighteen nucleotides in telomere ends could be elongated in the presence of an extract made from *Tetrahymena* macronuclei at the stage of minichromosome formation. They called the responsible enzyme telomerase. The puzzle of where the template for adding the DNA repeats to telomere ends was solved when Blackburn and Greider demonstrated that telomerase activity was abolished if the extracts were treated with an enzyme that destroyed RNA. They concluded that telomerase is an RNP.

As an independent researcher at the Cold Spring Harbor Laboratory, Greider isolated the *Tetrahymena* gene encoding the essential RNA and confirmed that telomerase uses the product of this gene, an RNA chain of 159 nucleotides, to produce telomeres. A very simple sequence in this RNA, containing one and a half complements to the DNA repeats of CAACCCAA, was shown to be the template for addition of many repeats to the natural ends of chromosomal DNA, presumably via a slipping and extending mechanism. In the following years, genes en-

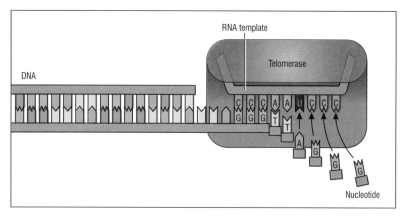

Fig. 3.7 Telomeres are formed by addition of repeated sequences to chromosome ends via reverse transcription. The template for repeated additions is a short sequence in an RNA molecule (green) that is an integral component of the enzyme telomerase.

coding the protein that catalyzes telomere extension in the telomerase RNPs were cloned from several organisms. As these proteins are closely related to the reverse transcriptases of retroviruses and retrotransposons, they were called telomerase reverse transcriptases (TERTs).

In humans, telomeres are made up of numerous repeats of the nucleotide sequence TTAGGG, and the 451-nucleotide telomerase RNA (TERC) includes the short, 1½ complementary templating sequence AAUCCCAAU. Our germ cells are rich in telomerase, but the amounts are low in the tissues of the body where most cells do not have to undergo numerous duplications. Cancer cells are a most unfortunate exception. In these malignant cells, telomerase production is reactivated, allowing the cells to multiply indefinitely. In recognition of the broad scientific and medical implications of their discoveries, Elizabeth Blackburn, Carol Greider, and Jack Szostak were awarded the 2009 Nobel Prize in Physiology and Medicine for determining "how chromosomes are protected by telomeres and the enzyme telomerase."

In nature, as elsewhere, there are often exceptions to rules. As it turns out, not all telomeres are created equally. In the fruit fly, *Drosophila melanogaster*, two families of non-LTR retrotransposons maintain telomere length by occasional jumps into the tips of chromosomes. This amazing mechanism provides yet another solution to the chromosome end problem via reverse transcription.

Coda

The shared features of genomes and the conserved biochemical mechanisms by which genes are expressed provide modern support to the conclusion that all living things on Earth are descendants of a common ancestor. As noted rather poetically by Charles Darwin in his *Origin of Species:* "Therefore I should infer from analogy that probably all the organic beings which have ever lived on this earth have descended from some one primordial form, into which life was first breathed."[4]

The origins of the last universal common ancestor (LUCA) of all cellular organisms, mobile genetic elements, and viruses are presumed to be intertwined, even though it is not clear to evolutionary biologists whether they coevolved or if one came first when life first arose. In the primordial world, RNA chains are now generally believed to have served as genomes and enzymes, most likely for eons, until proteins were recruited to enhance stability and efficiency. It is remarkable that examples of catalytic RNAs can still be found in today's world and, as elaborated in this chapter, many ribonucleoprotein complexes, RNPs of likely ancient origin, provide vital functions to extant life forms. On the other hand, RNA genomes have persisted only among some contemporary viruses, including the retroviruses.

As we have seen, discovery of the retroviral reverse transcriptase provided a framework for understanding the transition from RNA to DNA worlds. Phylogenetic relationships indicate that the gene encoding this enzyme was likely acquired from an ancient RNA transposable element at some point in the early evolution of retroviruses. The ability of these viruses to be transmitted from one host to another facilitated their broad distribution among members in the tree of life. The continuing quest to determine the origin of life, as well as an appreciation of the importance of RNA genomes and reverse transcription in evolution, has illuminated our view of the biosphere, in which the retroviruses have served as most valued beacons.

Retroviruses and Evolution

Even before Howard Temin proposed his revolutionary provirus hypothesis, there were hints that retroviral DNA could be included in the hereditary legacy of vertebrates. The signs were first observed in agricultural animals. When infections by a leukemia-inducing retrovirus related to the Rous sarcoma virus began to plague poultry farmers in the early 1960s, efforts were made to screen flocks for the known agent, called avian leucosis virus (ALV). The first method, described in 1964, used antibodies produced in hamsters with tumors induced by injection of the related Rous sarcoma virus. These antibodies were found to react not only with the injected virus but also with numerous stains of ALVs. It was surmised the antibodies must recognize a region common in proteins of these viruses and, therefore, was designated the "group-specific" antigen, or Gag. (We now know that this antigen resides in the capsid protein of these viruses, CA in the "gag" region as illustrated in Figure 2.4.) The first unexpected finding was certain chicken strains testing positive for Gag, even though they were known to be uninfected. How could that be? The initial thought was that the test was not sufficiently specific. However, the same investigators discovered that virus-like particles, as well as Gag-related antigens, were produced in certain tissues of the same uninfected chickens.

At about the same time, L. N. (Jim) Payne and Roger Chubb, at the Houghton Poultry Station in Britain, showed that production of the Gag-related antigen was inherited in Mendelian fashion in crosses between Gag-positive and Gag-negative chickens. Was it possible that a sequence similar to the viral Gag antigen was present in a normal cellular protein? Or could it be a Gag protein produced by some sort of latent virus?

These perplexing reports were surfacing when Robin Weiss was working on his doctoral research in Britain and pondering his own puzzling observations. Weiss had been culturing cells from embryos of various chicken lines infected with a defective strain of the Rous sarcoma virus that lacked an *env* gene. Because the virus was therefore unable to produce the viral spike protein, there should have been no infectious progeny issuing from these cells. To Weiss's astonishment, infectious virus particles *were* produced in some of his cultures, a phenomenon he reported in 1967. A similar result was described by another virologist in California, the Czech American Peter Vogt, following infection of cells cultured from Japanese quails. Weiss then discovered that the progeny viruses produced in his chicken cell cultures had unexpected host cell specificity, as if they possessed a novel Env-encoded protein. These observations, together with the Gag studies described above, led Weiss to suggest that Env protein might be produced from a retroviral genome integrated in the DNA of some chickens. His initial 1968 manuscript proposing this scenario was promptly rejected. Because reverse transcriptase had not yet been discovered, almost no one believed Temin's provirus hypothesis. One reviewer asserted that his interpretation was "impossible." However, with encouragement from Temin, Weiss submitted his manuscript to another journal, where his results and interpretation were published in 1969. Two years later, in experiments with inbred lines of chicken, Weiss and Payne showed that Env complementation and Gag-antigen expression are inherited together and, indeed, segregate as a single genetic trait in chickens. This was the first clear evidence for the existence of endogenous proviruses, retroviral-derived sequences that are an integral component of an animal's genome.

As is often the case in science, investigators in a number of laboratories who studied mice soon came up with results that were similar to those in the avian system. In 1969, researchers at the National Institutes of Health (NIH) reported that murine leukemia virus (MLV) particles were released "spontaneously" from uninfected cultures of mouse embryo cells. Another group found that production of a retrovirus that induced mammary tumors in mice (MMTVs) is an inherited genetic trait. Gradually, the "impossible" became the "accepted." Recognition of endogenous retroviral sequences in mice and chickens prompted many researchers to search for and identify similar elements in the genomes of other animals, including primates. Recognition of endogenous retroviruses in the human genome actually began with studies of DNA from a monkey.

Malcolm Martin and his colleagues at the NIH speculated that endogenous retroviruses in primate DNA might be detected by virtue of some similarity to those in mice, because viruses in the same genus as MLV (*Gammaretrovirus*) had been detected in some primate species. They began their search using nucleic acid hybridization, a popular technique at the time. The method is based on knowledge that single strands of DNA with complementary sequences will bind to one another, and under certain conditions, the interaction of even partially complementary sequences can be detected. Martin and colleagues prepared a "probe" comprising single strands of radioactive DNA from MLV and found that it bound fragments of DNA from an African green monkey. The probe was then used to identify a cloned fragment of monkey DNA containing a full-length, endogenous provirus with protein-coding sequences related to MLV. Although they could not detect hybridization of their MLV probe to DNA from human cells, they speculated a probe made from the endogenous monkey retroviral sequences might reveal the presence of human endogenous retroviruses. In 1981, they reported the detection of what appeared to be several human endogenous retroviral sequences with this probe, as well as their success in isolating the first human endogenous retrovirus clone.

Thanks to the development of next-generation DNA sequencing methods at the turn of this century and the resulting expansion of databases, we now know that endogenous retroviral sequences comprise a substantial proportion of the genomes of all vertebrates. In humans, endogenous retroviral sequences are distributed in an astounding 700,000 distinct sites in all forty-six chromosomes, representing about 8 percent of our genome. Many of these sequences are remnants from retroviral invasions of germline cells that occurred millions of years ago, but similar events have continued throughout evolutionary time.

One method to determine the timeline of insertion is based on an understanding of retroviral DNA synthesis and integration mechanisms. During reverse transcription, sequences from both ends of the

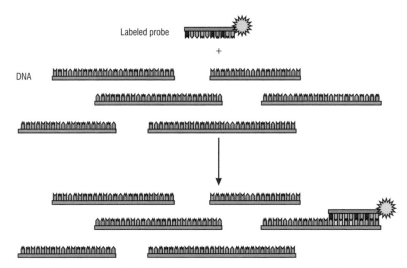

Fig. 4.1 Nucleic acid hybridization. Binding via complementary base pairing of a fragment of nucleic acid that is radioactively labeled (identified by the gold ball) can be used to identify and separate homologous sequences from a large number of unrelated nucleic acid fragments.

viral RNA are copied twice, to form the long terminal repeats (LTRs) flanking viral DNA (illustrated in Figure 2.3). Consequently, the sequences of these LTRs are identical at the time of invasion. Once embedded in the host DNA, each LTR will be subjected *independently* to mutations occurring randomly in the host genome, a result of random copying mistakes during DNA synthesis or various environmental insults. The frequency with which such mutations occur over time has been calculated for the cells of many organisms, including humans. The age of an endogenous retrovirus can be estimated by determining how many differences exist in the sequences of one LTR compared to the other. However, the most common method for assessing the age of endogenous retroviruses is based on estimates from evolutionary trees. The presence of the same endogenous provirus at identical (orthologous) locations in the genomes of two or more host species implies that the original invasion occurred in the DNA of a shared ancestor, prior to the date in which these species diverged from one another. Consequently, the provirus must have existed before such divergence was known to have occurred, in many cases millions of years ago.

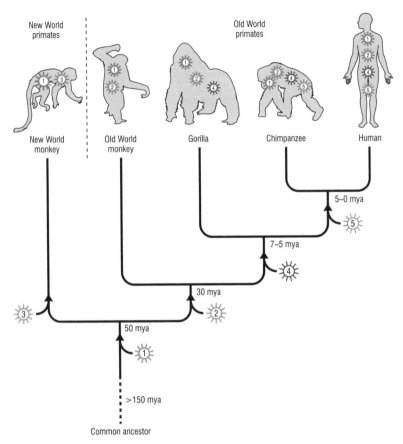

Fig. 4.2 Estimating the age of endogenous retroviruses from an evolutionary tree. Evolutionary trees approximate dates for the diversion of related species shown as millions of years ago (mya) for some primates. Species that share related endogenous retroviral sequences in analogous positions in their genomes are assumed to have inherited these sequences from a common ancestor infected by the retrovirus before the species diverged from one another. In the example shown, virus 1 is the oldest in the human genome: it is also present in the New World (Central and South America) monkey genome, from which all the New World primates (including humans) diverged over 50 million years ago. Because humans and chimpanzees share only virus 5, it can be considered the youngest human endogenous virus, acquired by a common ancestor of these species before they diverged, no more than about 5 million years ago.

Endogenous Retroviruses in Human DNA

In the decades since their discovery, more than fifty distinct endogenous retrovirus "families" have been identified in human DNA. Although first identified in humans, and therefore called *human* endogenous retroviruses (HERVs), they are believed to represent about 200 separate germline invasions occurring over millions of years, well before humans inhabited the Earth. The HERV-L family is among the oldest, estimated to have invaded the genome of an ancestor of all vertebrates about 100 million years ago. Based on the sequences of its *pol* gene and absence of an *env* gene, it has been suggested that HERV-L may be an evolutionary intermediate between LTR-retrotransposons and infectious retroviruses (see Figure 3.5). Other HERVs appear to have been acquired in bursts. For example, one group is derived from numerous retroviruses that invaded an ancestor of primates wandering the Earth some 45 to 30 million years ago. A few HERVs, such as some HERV-K members, are fairly "modern"; they are found only in the human genome, likely acquired fewer than a million years ago. As noted by Robin Weiss, "If Charles Darwin reappeared today, he might be surprised to learn that humans are descended from viruses as well as from apes."[1]

Considering the eons of time over which they existed, it is astounding that the vertebrate endogenous retroviruses can be classified according to the similarity of their *pol* genes with the eight genera of retroviruses circulating as infectious agents today (described in Table 2.1). This property indicates that retroviruses have existed close to their modern form for hundreds of millions of years. Although numerous, identified HERVs fall into only three classes: most are related to the betaretroviruses (like MMTVs), the gammaretroviruses (like MLV), or the spumaviruses, which, as noted in Chapter 2, have no known pathology.

As was apparent during the initial discovery of endogenous retroviruses in chickens and mice, some inherited proviruses can give rise to progeny virus particles that are able to infect other cells. Judging from the numerous copies of some endogenous proviruses in animal genomes, this was likely the case for the early invading retroviruses. However (perhaps fortunately), none of the HERVs identified in our genome to date have retained this potential. They all lost the ability to produce infectious particles prior to the divergence of humans and chimpanzees, about 6 million years ago. All now contain mutations or deletions preventing production of progeny. Indeed, in numerous cases, everything except an LTR has been deleted. These "solo" LTRs, which range in length from about three hundred to more than a thousand base pairs, mark the sites in which an endogenous retrovirus had integrated but where viral gene sequences were later excised via recombination between the homologous, flanking LTRs. Such recombination events have not been rare; human DNA contains ten times more solo LTRs than full- or partial-length endogenous proviruses. However, the lack of replication-competence and even of functional genes does not equate to a lack of impact. Several clear examples attest to the ability of HERV sequences to have a profound influence on human evolution and physiology.

Retroviral Sequences Are Hot Spots for Recombination Increasing the potential for DNA rearrangement is one way in which HERV-derived sequences and the other numerous retroelements in our genome affect human evolution and physiology. This feature, predicted by Barbara McClintock more than half a century ago, derives solely from the opportunity for recombination to occur between related retroelements embedded at diverse locations in host cell DNA. Recombination between two such elements embedded in the same orientation in a chromosome will lead to deletion of the cellular DNA that lies between them. If the elements are in opposing orientations, their recombination will lead to inversion of the intervening cellular DNA. Interactions between two (or more) related retroelements present in both copies of a particular chromosome offer additional opportunities for DNA rearrangement. For example, recombination between the first element in one chromosome and a second, downstream element in the other chromosome produces two distinct products: one product is a chromosome containing a single, recombined element but *missing* host DNA initially located between the first and second elements. The second product is a chromosome containing a reciprocally recombined ele-

(A) Deletions and inversions of host DNA via recombination between related HERVs on a single chromosome

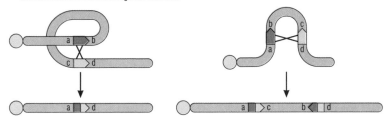

(B) Deletions and duplication of host DNA following recombination between HERVs in misaligned chromosomes

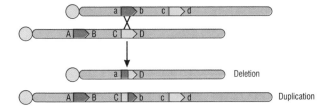

Fig. 4.3 Endogenous retroviral DNA recombination induces host gene shuffling. Recombination of endogenous retroviral sequences in a single chromosome will cause deletion or inversion of host genes, depending on their orientation (A). Misalignment and recombination between retroviral sequences in homologous chromosomes can lead to deletion of intervening host sequences in one chromosome and their duplication in the other (B).

ment flanked by *duplications* of the cellular DNA and the original first and second elements.

Loss of essential genes lying between related HERVs in the Y chromosome was deduced to be the cause of certain cases of male infertility. In theory, such deletions might result from recombination between the two flanking HERVs in a single chromosome. Evidence for this type of HERV-mediated genome reshuffling during primate evolution has been reported. Alternatively, HERVs could represent one product of recombination between two chromosomes, as described above. Direct evidence for the latter was reported in 2003 from a genetic study of men in southern Europe.[2] The Y chromosomes in 2 of 1,200 individuals surveyed were found to have duplications of the essential genes (flanked by the original HERVs) on either side of a recombinant HERV—an arrangement predicted for the second product of recombination between two chromosomes as described above. The investigators surmised that this recombination event probably occurred via misalignment of Y chromosome pairs in a sperm-producing cell of paternal forbearers. No detrimental effect was observed in the fertility of men who inherited the duplication, but the increased number of HERV and gene copies in their Y chromosome could promote additional rearrangements in the future.

Other types of genome contortions can be envisioned—from recombination between different chromosomes or within particular chromosomes in which related retrotransposable elements are embedded. The numerous copies of HERVs and other retroelements in the major histocompatibility complex (MHC) in human DNA are responsible for promoting the multiple DNA duplications and rearrangements that have driven the evolution of our immune system. These dynamic changes have produced a large repertoire of sequences, which allow formation of antibodies that attack foreign agents. We owe our ability to fight infection and disease not only to the transposable element from which the MHC was derived but also to the DNA shuffling promoted by resident HERVs and retroelements. In contrast to this beneficial example, the genetic defect associated with the Y chromosome deletion described above illustrates the essential "yin / yang" character of such elements, which can be both helpful and detrimental to their host organism. As described below, there is ample evidence from genomic studies that the genetic plasticity induced by endogenous retroviral sequences has been a driving force in evolution.

Regulatory Sequences May Be Purloined HERVs have affected human evolution and physiology through the introduction of novel regulatory units. The retroviral LTRs are made up entirely of regulatory signals. They include "promoters," short sequences that recruit host proteins required to initiate messenger RNA (mRNA) production, as well as signals that tell mRNA synthesis to terminate. Other LTR regulatory sequences, called "enhancers," are binding sites for proteins that can induce an upsurge in mRNA synthesis not only from the LTR in which they reside but also from cellular promoters located thousands of base

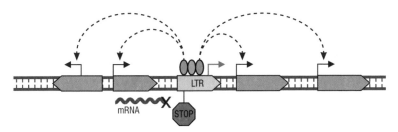

Fig. 4.4 Regulatory sequences in the LTRs of endogenous retroviruses can affect host gene expression. Host genes are represented as large blue arrows, indicating the direction of their transcription from the host DNA. Retroviral LTRs contain sequences that may promote transcription (red bent arrow) of host genes that are downstream or cause termination (via stop signal) of mRNA synthesis by host genes that are upstream. Enhancer elements in the LTRs can recruit host proteins (green ovals) capable of causing an upsurge in mRNA synthesis of host genes in either direction (dotted lines), some of which may be far away but brought close to the LTR by DNA bending.

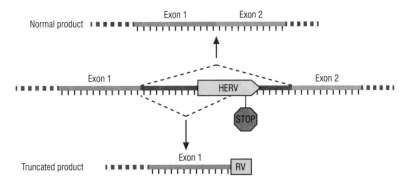

Fig. 4.5 Splicing signals in HERVs can lead to altered cell messenger RNA production and abnormal protein formation. When a HERV is integrated within the intron of a host gene, RNA synthesized by this gene (shown in the center) carries not only the splicing signals normally used to join exon 1 to exon 2 in production of mRNA (as shown at the top) but also splicing signals within the HERV. If the HERV signals are used (as shown at the bottom), an aberrant mRNA will be produced containing exon 1 joined to sequences corresponding to the end of the HERV. Translation of this truncated mRNA will produce a protein lacking information in exon 2 and containing a new end from sequences encoded in the HERV.

pairs upstream or downstream of the LTR, through protein interactions. Some LTRs contain signals that respond to hormones or other regulators present in only certain tissues of the body.

The protein-coding portion of endogenous proviruses also contains regulatory signals, sequences that define the ends of exons and control splicing of RNA transcripts. Use of such signals in HERVs that are located in introns or within the exons of cellular genes can lead to formation of truncated or altered mRNAs, which will then code for proteins that contain deletions or newly acquired sequences. All of these features have the potential to impinge on our genetic heritage.

Upon initial invasion, retroviral DNA can be integrated almost anywhere, in either direction, within or between host cell genes. Should proviruses disrupt essential host genes or regulatory DNA sequences, they will be lost because cells suffering such a catastrophe will not survive. Not surprisingly, most HERV-related sequences in the human genome are located outside of genes or within the introns separating protein-coding sequences in genes. As part of what has been described as a continuing evolutionary "arms race" between infectious agents and their hosts, numerous cellular, antiviral defensive measures have arisen. One such measure comprises silencing of integrated proviruses by chemical modification of cytosine nucleotides in their DNA and of amino acids in the bound cellular histones, a process known as epigenetic regulation (*epi* = "on top of" in Greek). Infected cells will produce progeny viruses until, in the fullness of evolutionary time, such silencing prevails. However, in some cases, HERV-related regulatory sequences have been prodded out of such dormancy to provide new opportunities for altering the composition or regulation of host genes. Furthermore, because the regulatory functions of some LTRs are amplified in particular cells or tissues of the body, the products of host genes under their control may appear at times or in places where they do not ordinarily exist.

Genome surveys suggest that some human gene transcription is driven or enhanced by LTR promoters. Several documented cases illustrate how such activities have affected human physiology and evolution. For example, DNA from an endogenous retrovirus that entered the human lineage 13 to 23 million years ago was inserted just upstream of an extra copy of the gene encoding amylase. Amylase is a digestive enzyme normally produced in the pancreas. The LTR now controlling the amylase copy contains signals for production of the enzyme in salivary glands as well as the pancreas. This HERV-facilitated expansion in tissue distribution was not a trivial event. It allowed our ancestors to thrive on a diet containing starch and eventually to make the transition from hunter-gatherers to master farmers. The LTR of another HERV regulates a cluster of genes encoding the globin part of the oxygen-carrying hemoglobin in our red blood cells. Enhancer and promoter signals in this LTR are critical for modulating the switch from production of the fetal to the adult form of globin during human development. Other examples of HERV-regulated, tissue-specific expression of human genes have been described in liver, colon, testes, and the placenta.

RNA splicing signals in HERVs and termination sequences in LTRs also provide opportunities for altering host gene expression. Utilization of viral mRNA-processing sequences can lead to production of truncated or elongated host proteins, as well as new amino acid combinations. For instance, an intron-embedded LTR has been shown to provide alternative splice sites in the human gene encoding the receptor protein for leptin. Leptin is a hormone secreted by fat cells in the body. Because it affects many physiological functions, including body weight, production of sex hormones, and blood cell production and activity, its concentration in the body must be controlled. The LTR-induced alternative splicing results in synthesis of a protein shorter than the normal leptin receptor and includes a retrovirus-derived amino acid sequence at its end. Although the short protein lacks normal functions, it can still bind leptin. It is thought therefore that the short protein may normally play a role in reducing the concentration of leptin in the bloodstream. These few examples of the impact of HERV regulatory signals on human physiology illustrate how exploitation of such transposable elements has influenced evolution of the animal species on our planet.

Retroviral Genes Can Be "Domesticated" Perhaps the most striking examples of the impact of endogenous retrovirus on evolution of mammalian physiology are in development of our reproductive systems. If it were not for inheritance of certain retroviral genes, we and other placental mammals might be hatching our young from eggs, like birds. In humans, the placenta is a "pancake"-shaped structure composed of cells from both mother and fetus. It is embedded into the wall of the uterus and connected to the fetus via the umbilical cord. Formation of the human placenta begins about five days after fertilization, when a layer of specialized cells called trophoblasts surrounds the small, inner mass of cells destined to become the fetus. Cells derived from this layer invade the uterine wall, where they fuse together to form multinucleated cells called syncytiotrophoblasts. The syncytiotrophoblasts provide a critical barrier between the mother and developing fetus. Because the fetus contains paternal gene products, it would be recognized as "foreign" and be destroyed, if accessible to the mother's immune system. The placenta shields the fetus from such danger. Nutrients and hormones supplied by the mother pass through the barrier to nourish the fetus as fetal waste products are removed. In a somewhat perverse way, one might think of the fetus as an invading pathogen, for which formation of the placenta is an effective survival strategy. In actuality, ancient retroviral invader genes have been coopted by their host to produce this protective barrier, allowing a fetus to develop safely within its mother's womb.

Retrovirus particles were detected in normal placenta of primates (including humans) and other animals in the 1970s, first with the electron microscope and later by using biochemical methods. In the following decades, a growing appreciation of the properties of retroviral proteins led some virologists to propose that the particles may play some role in formation of the placenta. By that time, it was known that retroviruses gain access to host cells with the help of their envelope (Env) proteins, which not only bind to cellular receptors but also possess "fusogenic" sequences that promote the merging of viral and cell membranes required for virus entry (illustrated in Figure 2.5). Other studies revealed that the retroviral Env protein also contains sequences that can suppress the host's immune response. But the full story of how retroviruses figure in the formation of the placenta was not revealed until the turn of this century.

In the year 2000, two separate research groups reported that a protein made abundantly in placental cells is the product of the *env* gene produced by a newly discovered human endogenous retrovirus, a betaretrovirus named HERV-W. Investigators in Francois Mallet's laboratory in Lyon, France, who first discovered the HERV-W family in 1999, noticed that *env* mRNA of this endogenous virus is produced only in placenta. After preparing a DNA clone from this mRNA, they discovered that the viral protein expressed by the cloned gene caused cultured cells to fuse with one another. As the HERV-W family is estimated to have invaded the primate lineage between 20 and 30 million years ago, these investigators reasoned that this particular *env* gene would not have survived debilitating mutations if it did not provide some benefit to the host. They proposed this function to be the cell-cell fusion reaction, the quintessential feature of placenta morphogenesis.

The second group of investigators, led by John McCoy in Cambridge, Massachusetts, happened upon a gene that they called "syncytin" while screening for human genes that encode novel secreted proteins. They made the connection between syncytin and retroviral Env proteins from a search of databases. Although they detected closely related sequences on eight separate human chromosomes, only the HERV-W insertion in chromosome 7 contained an intact *env* gene sequence identical to the cloned syncytin gene. In further tests, they also showed that the syncytin protein could mediate cell-cell fusion, as would be predicted for a functional role in placenta formation.

Within a few years, the Env proteins of two other HERV families were discovered to affect placental function. One of these, HERV-FRD, located on chromosome 6, invaded the primate lineage even before HERV-W more than 40 million years ago. HERV-FRD encodes a fusogenic Env protein made in cells of the trophoblast. This protein is called syncytin-2, to distinguish it from the HERV-W product, now known as syncytin-1. The second HERV, called ERV-3, is located on chromosome 7 and is expressed in the placenta, like HERV-W. Although its Env protein is not fusogenic, it does have immunosuppressive properties and, therefore, is likely to contribute to maternal tolerance of the fetus. In the past decade, many additional endogenous retroviruses have been shown to play a role in human reproduction as a result of tissue-specific regulation of host proteins and hormone production.

As impressive as the story of retroviral *env* gene domestication is for the human placenta, it is even more astounding that the same strategy has been repeated *five* independent times during mammalian evolution—with the exploitation of different endogenous retroviral sequences. Beginning about 60 million years ago, fusogenic properties of diverse Env proteins have been pilfered for distinct placenta development in rodents (mice, rats, and squirrels), carnivores (cats and dogs), lagomorphs (hares and rabbits), and ruminants (cows and sheep). It was recently reported that an endogenous retroviral *env*-derived syncytin contributes to formation of a placenta-like structure in a species of lizard. This striking pattern of convergence is driven not only by the evolutionary benefit associated with this mode of reproduction but also by the genetic potential extant in endogenous retroviral sequences.

It seems likely that regulatory sequences of one of the best-studied (and "youngest") families, HERV-K, were also exploited for human reproduction. Our genomes contain over 2,500 HERV-K solo LTRs and approximately 100 full-length copies of this endogenous retrovirus. Although every provirus has debilitating mutations in two or more genes, HERV-K is the only endogenous retrovirus family that includes several full-length members. By using modern-day recombinant DNA technology, scientists have been able to resurrect an infectious version of HERV-K in the laboratory by combining the functional parts from different copies in our DNA (shades of Frankenstein!).

Most HERV-K sequences residing in our genome are not normally expressed, having been silenced by epigenetic modification. However, a group at Stanford University, led by Joanna Wysocka, recently found that noninfectious particles, as well as some HERV-K proteins, can be detected in normal human fetuses at certain early stages of development. Because one of the retroviral proteins is known to block infection by different classes of viruses, including the influenza virus, they speculate that HERV-K induction might protect early fetuses from infection. There are other reports of retroviral sequence expression at various stages in human reproduction and indications of how they contribute to the process. For example, regulation by LTR sequences in one HERV family allows embryonic stem cells to maintain their capacity to develop into different tissue and organs. The genes of another HERV family are expressed in several fetal tissues, where they are thought to contribute to differentiation.

A Possible Dark Side As beneficial as the evolutionary interactions between endogenous retroviruses and their host genomes have been, these invaders can also have a dark side. Disruption or loss of important genes is one example, exemplified in the Y chromosome deletions described earlier in this chapter. It is not hard to imagine how reactivation of some normally silenced LTRs or the unscheduled expression of endogenous retroviral proteins could trigger or exacerbate disease. Numerous studies hint at possible involvement of HERVs in human pathologies, especially those of the immune system, brain, and central nervous system, as well as in certain types of cancer.

Autoimmune diseases are characterized by the attack of an individual's immune system on normal cells in the body. In some cases, an antibody to a "foreign" protein may recognize and attack a normal protein in the body, a mechanism called molecular mimicry. It has been suggested that antibodies targeting some HERV proteins might possess this property. However, HERVs could also contribute to immune cell dysfunction via inappropriate regulation of host genes. Activation of several HERV families has been detected in the blood and tissues of patients with various autoimmune diseases including rheumatoid arthritis, systemic lupus erythematosus, and type 1 diabetes. Products from reactivation of normally quiescent HERVs have also been detected in patients with multiple sclerosis, schizophrenia, and other neurological disorders. Although these associations have generated a great deal of interest, it has not yet been possible to determine if HERV activation is a cause of these diseases or simply an effect of stress produced by the pathologies.

The idea that HERVs may be associated with cancer started with the discovery of noninfectious, virus-like particles budding from some human tumor cells. Endogenous retroviral proteins, as well as antibodies to these proteins, are also found in humans with cancer. For example, HERV-K products are detected in 50 to 80 percent of patients with germ cell tumors, and HERV-K–derived particles, viral proteins, and antibodies to them are found in approximately 20 percent of patients with melanoma. The Env protein is detected in two out of three primary breast cancers, and there is some evidence that the Env protein could contribute to tumor pathology. Cultured breast cancer cells that express Env stop dividing and die when treated with an antibody targeting this protein. Because HERV-K is a betaretrovirus like MMTV (which causes breast cancer in its host mouse species), it is tempting to draw analogies between the two. However, as with the associations described above, we don't yet know if HERVs are perpetrators, accomplices, or simply innocent bystanders in human disease. It is an important question to answer, though. If convicted, strategies to silence HERVs or ameliorate their effects might lead to new therapies.

Interactions with Other Viruses Infection with various types of circulating viruses is another source of stress that can nudge certain HERV sequences out of slumber. Infection with two important human pathogens, the Epstein-Barr virus (EBV) and human immunodeficiency virus type 1 (HIV), has been shown to induce HERV expression—with potentially negative consequences.

EBV, a large DNA virus that infects the antibody-producing B cells and some T cells of the immune system, is one of the most common viruses of humans (its only host). Most people are infected by this virus at some point in their lives. EBV causes infectious mononucleosis (the kissing disease) and other diseases, especially in people with compromised immune systems. Viruses in this family have an unusual lifestyle. Once a person is infected, the viral genome remains in his or her B cells in a quiescent state known as latency. In this state, viral DNA is replicated by host cell enzymes and only particular viral genes are expressed. The quiescent genome can be reactivated, and virus production resumed at a later time by subsequent stresses, including unrelated infections. However, this is rarely associated with clinical symptoms.

The suspicion that EBV played a role in the pathogenesis of multiple sclerosis first arose when some patient studies revealed that people infected with EBV were more likely to develop the disease. As most people are infected with EBV by the time they reach mid-adulthood, it is hard to know for sure whether the virus really plays a role. Nevertheless, because expression of EBV latent stage proteins causes activation of certain endogenous retroviral sequences, including HERV-W and certain HERV-K members, it has been suggested that EBV may contribute to pathogenesis by inducing production of syncytin-1 and immune cell–stimulating HERV-derived proteins. It is conceivable that HERV proteins could trigger a cascade of reactions, leading to destruction of the myelin sheath that protects nerve cells in the brain and spinal cord. But multiple sclerosis is a complicated disease, influenced both by patient genetics and environmental factors. Consequently, it has been difficult to determine the degree to which HERV expression may affect pathology.

HIV infection can also cause activation of a number of HERV sequences in cultured human cells. Furthermore, HERV gene expression is elevated in the serum of patients with AIDS and increases as the disease progresses. The significance of these observations in the pathology of AIDS remains to be determined. However, studies have shown that the Gag and Env proteins of some HERVs can be incorporated into HIV particles. Although particles containing mixtures of these proteins show weak infectivity, use of a HERV Gag protein could allow HIV lacking a functional Gag gene to be transmitted to an uninfected cell. Furthermore, because the Env protein of HERV binds to a different cell receptor than that of HIV, its incorporation into particles could allow HIV to enter cells that it would not normally infect. In patients with AIDS, HIV has been detected in some cells that do not express the major receptor for entry. Whether HERV proteins contribute to this type of virus spread is unknown, but the phenomenon has been observed in various animal systems and is a contributor to virus evolution.

Cross-Species Transmission

In 1973, Jay Levy at the University of California, San Francisco reported an unexpected occurrence. Tissues prepared from the strain of mice under study in his lab contained retrovirus particles that he and others had thought to be defective for replication, because they did not infect mouse cells. However, most surprisingly, these particles *could* infect the cells of other animal species, including rat and human cells. Because they were similar to the infectious MLV particles produced spontaneously by cells of other mouse strains, Levy concluded that these particles were derived from an endogenous variant of MLV in the mouse genome. He called this variant "xenotropic" (from *xenos,* the Greek word for "foreigner") to signify that the virus could only infect the cells of foreign animal species. He noted at the time that similar observations had been made with certain retroviruses of cats.

In the process of evolution, loss of the endogenous retroviruses' capacity to reinfect the cells of their host is likely to be a selected trait, due to its benefit to the animal species' survival. However, the capacity to infect other species is a distinct advantage for the survival of retroviral genomes. By cross-species spread, an endogenous retrovirus of one animal can become a circulating retrovirus in other species. Complementation of any provirus defects in the new species, as well as genetic recombination, can lead to the generation of new retroviral variants with potential for even broader transmission. Moreover, if the germline cells of the new host species are susceptible to infection, invasion of their genomes can lead to the establishment of similar endogenous proviruses in diverse landscapes of different species, an evolutionary safeguard against possible extinction of its original host. Judging from the sequence relationships among endogenous retroviruses in different species and similarities with currently circulating retroviruses, these types of events have been occurring for millions of years—and, as described below, persist to this day.

Invasion of the Koala Genome The cuddly koala bear is an iconic symbol for Australia. However, the survival of this extraordinary animal is now being challenged by a retrovirus. A brief 1988 veterinary journal report offered the first description of leukemia in a koala, caused by retrovirus infection.[3] When the retrovirus (called KoRV) was isolated more than a decade later, it was found to be a Gammaretrovirus, closely related to a retrovirus then circulating in gibbon apes. But koalas only exist in Australia, miles and oceans from any gibbon apes, so how could this be? Virologists speculate that both animals acquired the retrovirus in the distant past, via cross-species transmission of endogenous retroviral particles from a rodent species, most likely feral mice. The enormity of the problem for the koala population only became apparent when captive animals in Dreamland, a theme park in Queensland, Australia, started dying of leukemia and lymphomas.

A test of 200 koala bears in the northern Queensland area, both captive and wild, showed every animal contained KoRV sequences. It was concluded that the circulating virus originated from an active endogenous provirus. Traces of the virus were subsequently detected in specimens from this population that had been preserved in museums 150 years ago, indicating KoRV has been infecting these animals for at least that long. In 2006, researchers at the University of Queensland identified proviruses at numerous sites in the genomes of many different animals. However, they found only a few koalas had proviruses at common sites. Researchers concluded that the circulating, infectious KoRV was invading germline cells repeatedly in the population. This interpretation was confirmed by finding identical patterns of proviruses in three generations of related koalas and in the DNA of all cells (including the germ cells) from single individuals.

The next surprise came when koalas from other parts of Australia were examined. As tests were extended southward, increasingly smaller fractions of koalas were found to be infected with KoRV and to carry the endogenous sequences. The smallest number was found on Kangaroo Island, 16 miles off the southern coast, where koalas were introduced in the early 1900s. It seems clear, therefore, that colonization of the koala genome by KoRV is an "act in progress"—providing a unique, observable snapshot in evolution. In the years since their discovery, KoRVs seem to have lost some of their virulence, with fewer infected animals dying of leukemia. It is possible that given sufficient time, more benign variants of KoRV will predominate and, as with other endogenous retroviruses, the virus will no longer harm its host.

The Host-Virus Arms Race

"Arms race" is a metaphor for the relentless struggle between viruses and their hosts, in which adaptations on either side are met by countermeasures on the other, as both strive to survive in an ever-changing environment. As stated by the Red Queen to Alice, "Now you see, it takes all the running you can do, to keep in the same place."[4] According to Darwinian principles, the host-virus arms race is driven by random mutations (including sequence rearrangement) followed by selection of those affording the best advantage for propagation. It is easy to imagine how host genes associated with antiviral defense evolved in this way. Individuals with genes that were ineffective would die of infection and be eliminated from the population. On the other hand, any viral mutants able to evade the antiviral defense would flourish, again exerting selective pressure back on the host, in an ever-escalating competition.

In this continuing struggle, it may seem that host organisms would have an edge. Viruses need proteins encoded in many host genes for

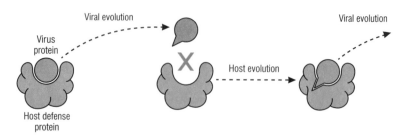

Fig. 4.6 The host-virus evolutionary arms race. Changes that lead to evasion of host defenses arise in rapidly evolving viral sequences. Hosts can survive infection if mutations arise that allow the host defense to engage the changed viral protein, initiating another round of both viral and host evolution.

propagation, and such dependence could afford numerous opportunities for changes that would inhibit the process. However, although viral genomes are limited in size, very large numbers of viral progeny are typically produced in each reproductive cycle, with many mutants among them. This property increases the probability for one or more of such genetic variants to be capable of circumventing such inhibition. In addition, retroviruses have also acquired unique ways to make use of host resources. They not only produce very large numbers when propagated as circulating viruses but can also make the most of their host genome when propagated as heritable, endogenous proviruses.

From a virus standpoint, it would make good sense to depend on those host proteins unlikely to be lost or change drastically by mutation. Regions in such proteins that are critical for function cannot be changed, such as those specifying the binding site for a hormone or the catalytic function of an important enzyme. Indeed, such strong sequence conservation makes it possible to identify reverse transcriptases across species, representing various stages in evolution. However, changes in other parts of proteins can often be tolerated without affecting essential function. Harmit Malik and his colleagues at the Fred Hutchinson Cancer Center in Seattle reasoned that such flexible regions would be just the place to search for molecular battlegrounds in the continuing host-virus arms race. They began by analyzing the sequence of a known antiviral gene in ten different primate species, representing about 30 million years of evolution. Although most parts of the gene were well conserved in all species tested, a few regions contained an unusually high number of changes that altered the encoded amino acids. Such changes are a sign of positive selection, consistent with the idea that these regions encoded sites of interaction between the antiviral protein and its viral targets. In further support of this interpretation, a single amino acid difference in one of these regions determined whether the antiviral protein was susceptible to a component of HIV. The approach of looking for such rapidly evolving regions led to identification of important points of contact between HIV and a number of antiviral cellular defense proteins. Sara Sawyer, a colleague of Malik, extended this strategy to pinpoint a battleground site shared by several distinct viral species but in this case in an essential, so-called housekeeping gene.

Located on the cell's outer surface, the transferrin receptor protein controls uptake of iron, a vital function for all living cells. This receptor is used by a number of viruses for entry into their host cells. When Sawyer and her students at the University of Texas compared sequences in the gene encoding this protein in rodent species, they found variations in a few codons. These variations were distant from the region required for iron transport, but they overlapped known binding sites for the Env protein of MMTV and the entry proteins of other viruses in both rodents and carnivores. Most convincingly, the investigators showed that expression of a receptor protein containing just three of the substitutions identified in this region was sufficient to prevent entry of MMTV into a host cell, while iron uptake was unperturbed. They concluded that repeated changes of particular amino acids are a signature of evolutionary adaptation in the arms race between host and viral genes.

In 2016, David Enard and colleagues in Dmitri Petrov's laboratory at Stanford reported results from their genetic analysis of more than one thousand host proteins known to interact with viruses at various stages of replication. Most were viruses that infect humans, and the majority of the host genes encoded retroviral (HIV)–interacting proteins. Nearly all of these proteins are associated with vital "housekeeping" functions, such as mRNA production, protein processing, and molecular transport; only 5 percent had any known antiviral activity. When the gene sequences of twenty-four mammalian species were compared, significantly more adaptive changes were observed in sequences specifying the proteins they identified as interacting with viruses than those that did not. An estimate of 30 percent of the codon changes in these genes occurred in response to viral contact. Signatures of evolution are not unexpected in proteins interacting with viruses, because viruses have been major predators in all branches of the tree of life as far back as we can trace. However, the enormity of viral impact on the composition of ordinary housekeeping genes *was* a surprise. It is hard to escape the conclusion that viruses are one of the most dominant drivers of evolutionary change.

A Continuing Saga

Although it has been half a century since the first endogenous retroviruses were discovered, the magnitude of their contributions to host genomes could not have been appreciated until large-scale DNA sequencing methods were developed. We now know that endogenous retroviral sequences comprise a substantial portion of the DNA of all eukaryotic organisms. In humans, these sequences are four times more abundant than those encoding all the proteins in our body. Endogenous retroviruses and the other retrotransposable elements in our DNA, first thought to be "junk" or leftover baggage from bygone days, have affected the evolution and physiology of their hosts in profound ways.

The latest news from genome prospectors is further enlightening. Not only have viruses driven the evolution of genes encoding antiviral defenses, but they have also shaped those that control the basic functions required by all living cells. Yet these are still early days. Sequence databases continue to expand, as do tools for data mining and for analyzing the effects of endogenous viruses and their retroelement cousins. We may have discovered only a fraction of their impact on evolution.

Revealing the Genetic Basis of Cancer

As we learned from the story of their discovery (Chapter 1), retroviruses first caught the attention of biomedical researchers because they caused cancers in various animal species. The idea that these viruses might hold a key to understanding the origin of these diseases was a powerful stimulus for continued investigation. Early researchers found oncogenic (cancer-causing) retroviruses could be classified into two main groups. The first group, originally called the *acutely transforming* retroviruses, comprised highly carcinogenic agents that cause malignancies in nearly all infected animals within just a few days. Some of these retroviruses, for which the avian Rous sarcoma virus was the prototype, also produced colonies of transformed cells in culture, first demonstrated in the focus assay developed by Howard Temin and Harry Rubin. Members of the second group, the *nonacutely transforming* retroviruses, were less carcinogenic. Not all infected animals developed cancer, and it took weeks or even months before malignancies appeared. Retroviruses in the second group do not transform cells in culture, but they can be propagated in such cells. With few exceptions, the infected cells continue to multiply while appearing normal even as they produce prodigious numbers of progeny virus particles. Unraveling the mechanisms by which these two distinct groups of oncogenic retroviruses cause cancers in susceptible hosts provided the first key insights into the genetic basis of these diseases.

Rous Sarcoma Virus and the *src* Paradigm

In the early 1970s, following discovery of reverse transcriptase and confirmation that retrovirus propagation depends on integration of its DNA into that of its host cell, Robert Huebner and George Todaro at the National Institutes of Health (NIH) reported that retrovirus particles are actually present in the cells of many, and perhaps most, vertebrates. They also observed that retroviral gene expression is associated with tumor incidence in several strains of mice. Huebner and Todaro theorized that endogenous retroviruses, which may be transmitted from animal to progeny and from cell to cell, could be responsible for transforming a normal cell into a tumor cell. For scientists interested in understanding how retroviruses caused cancers, the acutely transforming Rous sarcoma virus of chickens offered unique technical advantages. Use of the focus assay demonstrated that a single particle of this retrovirus is sufficient to transform an infected cell. Equally important, this same assay allowed identification of informative mutants. Much of the pioneering work with the Rous sarcoma virus can be traced back to the UC Berkeley laboratory of Harry Rubin, an inspiring mentor who "set the direction for all of us and gave us the essential tools to be successful and effective scientists."[1]

In 1970, Steve Martin, a research fellow working in Rubin's laboratory, reported the isolation of a unique mutant of this avian sarcoma virus. The mutant virus infected cells and produced infectious progeny at either 36°C or 40.5°C, but foci were only formed at the lower temperature. The mutant was temperature sensitive solely for transformation. Martin's groundbreaking report was the first clear evidence that some portion of the sarcoma virus genome was required for transformation, but this genetic information was separate from that required for viral replication.

The existence of separate genes for retrovirus propagation and cellular transformation had been suggested in earlier reports of avian sarcoma virus strains with properties opposite to those of Martin's mutant; they could form foci but were defective in propagation. Production of infectious progeny from these transforming viruses required coinfection with a replication-competent "helper" virus. Peter Vogt, a former Rubin trainee, showed that purified stocks of the avian sarcoma virus would give rise *spontaneously* to a significant percentage of progeny that propagated normally yet were not able to transform cells. Vogt's results were consistent with earlier observations by another Rubin trainee, Hidesaburo (Saburo) Hanafusa. Hanafusa had reported that preparations of the Rous sarcoma viruses were invariably associated with a substantial proportion of virus particles that did not form foci on cultured cells. However, these particles did produce leukemias when injected into chickens. These Rous-associated, avian leucosis viruses are known as ALVs.

The next critical pieces of the puzzle came from direct analysis of the sarcoma virus RNA genomes and the genomes of their related ALVs. Peter Duesberg, a young German chemist working at UC Berkeley, had been characterizing the RNA of other viruses. Duesberg and Vogt decided to combine their expertise for a closer look at the RNA in these avian retroviruses. They reasoned that the genomic structures of transformation-competent and transformation-defective retroviruses might differ, as the latter had lost the ability for focus formation. Their analysis of RNA from isolated virus particles, published in 1970, revealed two types of RNA in these viruses, comprising a longer and a shorter strand.[2] Because the transformation-defective ALV particles contained only the shorter RNA, and sarcoma virus particles contained both types of RNA, Duesberg and Vogt proposed that the longer strand must include the transforming gene. To investigate the transforming component, they used a new analytical method in which fragments produced from the RNA genomes were separated by length and charge. Because the fragment pattern produced by each type of RNA is unique, the method was called "fingerprinting." Their analyses showed that fingerprints obtained with RNA from the avian sarcoma viruses and the transformation-defective ALVs were almost identical, except for one or two fragments only present in the sarcoma virus fingerprints. This was a major step forward, as transformation capacity could now be assigned to a distinct portion of RNA, presumed to encode a gene thereafter called *src* (for sarcoma). In subsequent studies Vogt, Duesberg, and their colleagues applied fingerprinting and other methods to show the transforming gene is located at the tail end of the viral genome. Their data, together with other researchers' genetic studies, established the order of genes in the Rous sarcoma virus to be *gag-pol-env-src*. The associated ALVs retain the same first three genes but lack *src*, which is deleted from the sarcoma virus with relatively high frequency. We now know this deletion occurs via recombination promoted by the presence of two short repeated sequences that flank *src*.

The attractiveness of the Rous sarcoma virus as a model to investigate the origins of cancer, as well as the excitement surrounding the discovery of reverse transcription, motivated two medically trained investigators with a shared passion for research. Newly arrived at the University of California, San Francisco in 1968, J. Michael (Mike) Bishop had set up his laboratory with the intention of initiating studies of the Rous sarcoma virus. He was joined shortly thereafter by Harold Varmus, fresh from an apprenticeship in a molecular biology laboratory at the NIH. To familiarize themselves with retrovirus biology, the two eager young investigators focused first on how and when retroviral DNA is made in infected cells. Their approach was to exploit the technique of nucleic acid hybridization (illustrated in Figure 4.1) with which Bishop had become quite proficient. Radioactive DNA probes were produced by reverse transcription of retroviral RNA and then used to monitor the synthesis of retroviral DNA in newly infected cells. The results of this work contributed to delineation of critical events in the early phase of retroviral propagation, when retroviral DNA is produced and eventually integrated into the genome of an infected cell (see Figure 2.5). Of equal importance, these experiments established the technical foundation for efforts to solve the mysteries of where *src* came from and, eventually, how *src* transforms cells.

Bishop and Varmus were members of a collaborative retrovirologist group in California that met periodically to share information and exchange ideas. After learning of Duesberg's and Vogt's analysis of RNA from the sarcoma and transformation-defective retroviruses at one of these meetings, Bishop and Varmus hypothesized it should be possible to obtain a nucleic acid probe specific for *src* sequences by a process of "subtractive hybridization." The strategy they developed started with preparation of radioactive DNA fragments representing the entire sarcoma viral RNA genome. This DNA was then purified and allowed to hybridize with RNA from the related ALV. The mixture was then applied to a column that retained any DNA base-paired with RNA, while

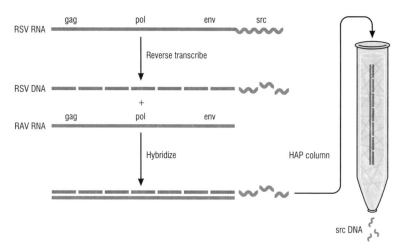

Fig. 5.1 Isolation of a *src* probe by subtractive hybridization. DNA fragments complementary to the RNA genome of the *src*-containing Rous sarcoma virus were prepared by using the retroviral reverse transcriptase and radioactive DNA nucleotides as substrates (blue dashes and red curves). The purified, radioactive DNA fragments were then hybridized with genomic RNA from a Rous-associated, avian leukosis virus that lacks the *src* gene. The mixture was then applied to a column that retained any DNA bound to RNA, allowing fragments containing the *src* DNA sequences to flow through.

releasing single-stranded, *src*-specific DNA molecules, which had no place to bind in the ALV RNA. This unique radioactive probe was now ready to reveal *src*'s origin. It seemed unlikely that *src* was an authentic retroviral gene because it was not required for retroviral propagation. But then, where did it come from?

The *src* probe was first tested for its ability to bind to DNA from uninfected chicken cells. A positive result suggested that *src* might have originated from host DNA, but enthusiasm was dampened by the fact that a probe made from the transformation-defective ALV also bound to this chicken DNA. Therefore, it seemed possible that the *src* probe was binding to copies of some unknown endogenous viral sequences. This idea was tested in hybridization experiments with DNA from evolutionarily distant avian species, including duck, quail, and even the Australian emu. Reported in a 1976 paper from their laboratory together with Peter Vogt, the results were riveting![3] Their *src* probe hybridized with sequences in each one of these avian DNAs, whereas an ALV probe showed little reaction. This result implied that part or all of the transforming gene of the Rous sarcoma viruses is derived from an evolutionarily conserved cellular gene, somehow "captured" from the genome of an infected chicken or a related avian species. Later studies from the Bishop / Varmus laboratory revealed that *src*-related sequences are present in virtually all multicellular organisms, from vertebrates down to the worms and sponges. In 1989, they were awarded the Nobel Prize in Physiology or Medicine "for their discovery of the cellular origin of retroviral oncogenes."

Following the revelation of *src*'s cellular origin and fueled by generous support from the NIH's Virus Cancer Program, discoveries in numerous laboratories added considerable depth to the story. The *src* oncogene was found to encode a distinct protein. In 1978, the viral protein, called v-Src, was revealed to be an enzyme that attaches phosphate to tyrosine residues in certain cellular proteins. Importantly, several cellular targets of v-Src were identified as membrane receptors responsible for receiving and transmitting signals from outside the cell, processes critical for normal cell function and proliferation. Such transmission is accomplished by relays of chemical reactions inside the cell, known as "signal transduction." Studies of the cellular counterpart of the viral oncogene, called c-Src, disclosed that this protein is normally inactive. The phosphorylating capacity of c-Src is only switched on when it is bound to a cell surface receptor actively engaged by some environmental signal. Activated c-Src is then able to modify target proteins required for a particular function or to initiate cell division. However, the *v-src* gene of the Rous sarcoma virus was corrupted by mutations that disabled the on / off switch. The resulting continuous production of this disabled protein in infected cells leads to inappropriate modification of receptor proteins, nonstop signaling, and, eventually, unfettered proliferation. The *v-src* of the acutely transforming avian sarcoma virus was the first retroviral oncogene shown to promote abnormal growth by disrupting the cellular signaling networks controlling cell growth, metabolism, and many other vital processes. But it was not the last.

With *v-src* as a paradigm, all acutely transforming retroviruses were considered potential carriers of cell-derived oncogenes and were known thereafter as *transducing retroviruses.* The next two decades were marked by a frenzy of activity as numerous other transducing retroviruses of birds, rodents, and other animals were mined for additional captured oncogenes and for insights into the mechanisms by which they caused various cancers. Remarkably, the Rous sarcoma virus proved to have the unique and fortuitous distinction of being the only member of this retrovirus group that can be propagated independently. All other trans-ducing retroviruses proved to be defective for viral replication. Having lost retroviral genes during oncogene capture, they can only produce progeny when coinfected with a helper retrovirus supplying the missing functions. Such mixed infections made it initially quite difficult to iden-tify the transforming components. One can only marvel at the good fortune of those who focused first on the Rous sarcoma virus as their experimental system.

One of the next retroviral oncogenes discovered would eventually be the source of equally profound insight into the genetic basis of cancer. Acquired by the transducing avian retrovirus called MC29, this oncogene causes lymphocytic leukemia and carcinomas in infected chickens.[4] Fingerprint analysis indicated that the MC29 genome in-cluded a nonviral sequence unrelated to *src.* The novel sequence was joined to *gag,* but other viral genes were absent. By the early 1980s, availability of DNA cloning and sequencing methods made it possible to characterize the transduced oncogene, called *v-myc* (for myeolosar-coma / carcinoma), as well as its cellular counterpart, *c-myc.* c-Myc is a nuclear protein that functions at the effector end of several signal trans-duction pathways and is a key regulator for expression of cellular genes required for normal cell growth and proliferation. Unlike *v-src,* *v-myc* contains no mutations affecting its function. However, the con-centration of the cellular counterpart, c-Myc, is tightly regulated in normal cells, and such control is lost when the gene is expressed in the context of a provirus. Increased production of Myc protein in MC29-infected cells leads to abnormal growth and the proliferation charac-teristic of malignancy.

In the ensuing years, many retroviral oncogenes were identified in the transducing retroviruses, some of which were picked up indepen-

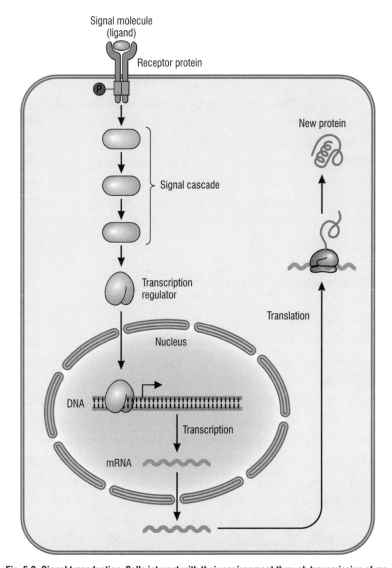

Signal molecule
(ligand)

Receptor protein

Signal cascade

Transcription
regulator

New protein

Translation

Nucleus

DNA

Transcription

mRNA

Fig. 5.2 Signal transduction. Cells interact with their environment through transmission of molecular signals from surface proteins, called receptors, which extend through their outer membranes. The receptors are stimulated by the lock-and-key binding of particular molecules called ligands (green oval). Changes resulting from this interaction are transmitted through a cascade of proteins by physical or chemical interactions, many of which involve the enzymatic addition or removal of phosphate. Nuclear proteins (such as c-Myc) respond to these changes by altering transcription of appropriate genes, from which new proteins may be formed by cytoplasmic ribosomes. Some new protein products may be excreted by the cells, thereby sending signals to others. This simplified diagram shows a single, hypothetical pathway. In reality, the many receptors and relay proteins in a cell form complex, interacting networks that control cell metabolism, growth, and proliferation. In multicellular organisms, signal transduction pathways have evolved to regulate cell communication.

dently several times from the genomes of the same or different animal species. The proteins they encode were found to impinge on many of the numerous steps in the transmission of critical intracellular signals. Transduced oncogenes included captured cellular genes for growth factors, their receptors, and enzymes functioning in chemical reactions required to transmit information from the cell surface to the nucleus for gene activation or control. While this work was in progress, researchers were also trying to discern how cancers are caused by the second group of retroviruses, nontransforming agents that lack oncogenes, and why it takes so much longer after infection for the disease to appear.

Insertional Mutagenesis and the *myc* Paradigm

By the 1980s, DNA sequencing and genetic studies revealed that integrated retroviral DNAs, the proviruses, comprise viral genes flanked by long terminal repeats (LTRs) (illustrated in Figure 2.4). Furthermore, although the "upstream" LTR controls expression of viral genes, the "downstream" LTR, at the other end of the genome, contains identical regulatory signals. With such powerful regulatory sequences at both ends, it occurred to some investigators that the downstream LTRs might, in some cases, be capable of promoting deregulated expression of neighboring cellular genes. If a provirus was inserted next to the cellular counterpart of one of the viral oncogenes, it might explain how nontransforming retroviruses could cause cancers. Moreover, because retroviral DNA is inserted almost randomly in the host genome, the low probability of an integration occurring in the vicinity of a potential oncogene could account for the length of time needed for cancers to develop in infected animals.

With no facile cell culture system for studying their biological effects, analysis of the oncogenic properties of the nontransforming retroviruses depended on selection of uniquely informative animal models. Here again, the avian system proved to be most useful, as the chronology of oncogenesis in ALV-infected chicks had been well documented. The critical experiments with this system were carried out independently, but simultaneously, by two sets of collaborators: the Bishop / Varmus group at UC San Francisco and William (Bill) Hayward at the Rockefeller in New York together with Susan Astrin at the Fox Chase Cancer Center in Philadelphia. In both studies, chicks were infected within the first week of hatching and followed for four to seven months, during which approximately 40 percent developed lymphomas derived from antibody-producing B cells.[5] When these tumors were isolated and their DNA analyzed, the tumors proved to be *clonal:* that is, each contained a population of cells in which ALV was integrated at the very same site, indicating that each tumor was derived from a single infected cell. In many cases, the proviruses in these tumors suffered deletions, resulting in loss of viral gene expression, but all retained at least one copy of an LTR. Most important, all of these integrations were mapped to within or near *c-myc,* the very same cellular gene captured by the transducing retrovirus MC29. Furthermore, the intact LTR in the ALV-induced tumors promoted levels of *c-myc* expression up to 100 times higher than normal. Although the exact integration site varied in each tumor, many proviruses were inserted within the first intron of the *c-myc* gene. Unexpectedly, however, some proviruses were located downstream of the *c-myc* gene, and several were in orientation opposite to the cellular gene.

Analysis of the relationship between the orientation of provirus insertion and *c-myc* expression in these avian tumors provided the first evidence for two distinct types of insertional mutagenesis. In the first type, *promoter insertion,* messenger RNA (mRNA) synthesis initiating from an adjacent LTR in the same orientation as the *c-myc* gene correlated with its vast overexpression. In many cases, such promotion was facilitated by deletions joining parts of the viral DNA to *c-myc.* These fusions placed the cellular gene under control of the viral LTR, thereby negating its normal regulation. The promoter insertion mechanism supported the ideas originally motivating these experiments. On the other hand, it did not explain how proviruses downstream of *c-myc,* or in the opposite orientation of *c-myc,* could cause overexpression of the cellular gene. This mystery was solved when the LTR was found to contain "enhancer" signals, sequences that bind cellular proteins that cause an increase in mRNA production, not only from the LTR but also from nearby cellular promoters, regardless of their orientation (illustrated in Figure 4.4). The second type of mutagenesis was therefore called *enhancer insertion.*

Discovery of the same cellular gene implicated in carcinogenesis by both transducing and nontransforming avian retroviruses was a truly unifying revelation. Moreover, the mechanism of promoter insertion by the nontransforming ALV immediately suggested pathways by which the transducing retroviruses might arise. For example, RNA produced by the fused virus + cell sequences in an ALV-infected tumor cell would be a plausible substrate for packaging into helper progeny particles together with the helper virus genome. On subsequent infection of a new host cell, copying from both helper and fused RNA during reverse transcription could produce a recombinant DNA product containing the cellular sequence flanked by two LTRs. This recombinant DNA could then be integrated into the host chromosome, producing a defective transducing provirus, ready to initiate tumor formation via the now captured oncogene, and to be propagated with assistance from its helper virus.

The *myc* story presented a most satisfying model for the origin of retrovirus-induced cancers in avian and other animal species. But the relevance of these discoveries to human disease was far from certain. Most types of human cancer did not appear to be contagious, and despite valiant efforts by researchers, there were no known human retroviruses of either the transducing or nontransforming type. As it turned out, investigations from an entirely different direction closed the conceptual gap between retroviral oncogenes and human cancer, with *c-myc* again in the spotlight.

The knowledge that integration of retroviruses could cause deregulation of cellular oncogenes in animals prompted collaborators Carlo Croce, at the Wistar Institute in Philadelphia, and Robert (Bob) Gallo, at the NIH, to wonder if the same phenomenon might be triggered in another way. Chromosomal abnormalities and rearrangements are characteristic of many human tumor cells, and it seemed possible that

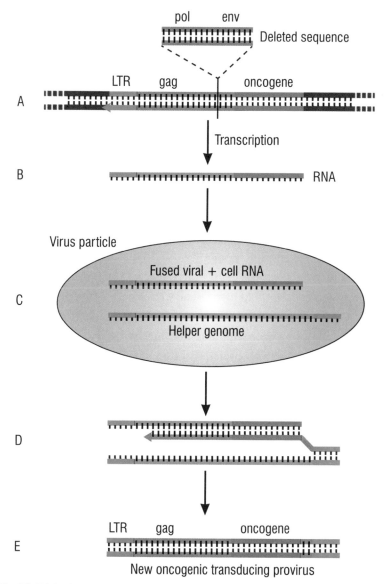

Fig. 5.3 **Origin of transducing retroviruses.** This diagram illustrates how a transducing retro-virus can arise following insertional mutagenesis by a nontransforming retrovirus. **(A)** An on-cogene is activated by a nontransforming virus via promoter insertion, following deletion of some viral and cellular sequences. **(B)** mRNA produced from the proviral LTR includes some viral as well as cellular oncogene sequences. **(C)** The chimeric mRNA is packaged into viral particles made by the helper virus, together with intact helper virus genomic RNA. **(D)** Upon infection of a new cell, reverse transcriptase copies from both the helper and chimeric RNA to produce double-stranded DNA that includes the oncogene and contains viral sequences at both ends. **(E)** The DNA product, comprising the oncogene embedded in viral sequences, is integrated into the host genome to form a transducing provirus.

such genetic reorganizations might also promote aberrant oncogene expression. Because increased levels of *c-myc* mRNA had been reported in various human tumors, they began by mapping the human gene to within a small region at the tip of human chromosome 8. This was an important clue, as chromosomal rearrangements that encompass this region were a known hallmark of the malignancy called Burkitt lymphoma, a fast-growing human tumor that originates in B cells and is associated with impaired immunity. In this disease, a reciprocal exchange between the tips of chromosome 8 and chromosome 14 (most common), 2, or 22 is seen consistently in tumor cells. In their 1982 report, Croce and Gallo noted the three chromosomes carry antibody genes, which are highly expressed in B cells.[6] They suggested that the rearrangements in Burkitt lymphoma cells might result in enhanced expression of a translocated *c-myc*. Later studies showed that the *c-myc* gene was, indeed, overexpressed in all tumor cells with chromosomal 8 to 14 translocations. Specific breakpoints for the DNA exchanges varied, ruling out a promoter insertion mechanism, as originally suspected. Instead, it was discovered that the translocations placed expression of the *c-myc* gene under control of strong enhancer sequences present in the adjacent antibody gene's promoter. These studies of Burkitt lymphoma provided the very first indication that cellular counterparts of retroviral oncogenes could contribute to human disease. Following this groundbreaking work, other examples of oncogene activation, via chromosomal rearrangement or by mutation, were identified in human cancers, consolidating the view of cancer as a genetic disease.

As a footnote to history, the genetic basis of cancer was actually predicted more than a half century earlier, long before the application of molecular methods and the confluence of diverse lines of investigation. In 1914, the German zoologist Theodor Boveri noted a close association between chromosome imbalances and several detrimental effects, including unchecked growth, in his experiments with sea urchin eggs. With astonishing prescience, Boveri connected these observations to the frequent chromosomal aberrations in cancer cells, predicting the existence of oncogenes, tumor suppressor genes, and the type of contributing environmental exposures (he called them poisons) such as radiation and viruses. Like many of the scientific pioneers, it took decades for his brilliant insights to be fully comprehended. It is now appreciated that all human cancers are associated with the activation of (usually multiple) oncogenes.

Discovery of an Oncogenic Human Retrovirus

The tragic loss of a younger sister to leukemia as a young boy affected Bob Gallo profoundly, kindling a desire to find a way to reduce the burden of this disease. After earning a medical degree and performing research at the University of Chicago, Gallo moved to the NIH in 1965 as a clinical associate, where he cared for patients and initiated a program of research focused on the biology of human blood cells. Following passage of the National Cancer Act and expansion of the NIH in 1971, he was appointed head of a newly formed Laboratory of Tumor Biology.

Blood cell malignancies are common human afflictions. It was possible, therefore, in the late 1960s to obtain sufficient quantities of some types of leukemia and normal immune cells to compare their properties. However, culture methods were still in their infancy when Gallo began his research, and human lymphocytes could not be maintained in culture for more than a few days without dying. To solve this problem, Gallo's group set out to find conditions in which human immune cells could be maintained outside of the body, taking clues from colleagues' success in identifying a substance that stimulated the growth of other cultured blood cells. In 1976, the Gallo lab reported that a protein secreted by normal T lymphocytes, later identified as interleukin-2 (IL-2), promoted the growth of these normal lymphocytes as well as some malignant counterparts in culture. This T cell–stimulating molecule was an essential component in the discovery of the first human retrovirus.

Despite the many successes with animal systems, decades of searching for human retroviruses had consistently met with failure or

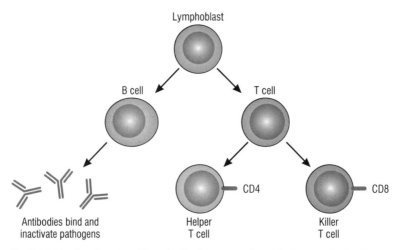

Fig. 5.4 Major lymphocytes of the adaptive immune system. The human adaptive immune system responds to invasion of viruses and other pathogens by mobilizing two important types of immune defense cells called lymphocytes. B cells are lymphocytes that develop in the bone marrow and produce antibodies that bind and inactivate pathogens carrying antigens that they recognize as "foreign." T cells are also formed in the bone marrow but mature in the thymus gland. Unique surface proteins, called CD4 or CD8, characterize two types of T cells. The CD4 protein facilitates binding to antibody-producing B cells. The CD4-bearing T cells are called "helper" cells because they enhance the production of antibodies by B cells. The CD8 protein facilitates binding of T cells to pathogen-infected cells, which are recognized by their production of foreign antigens. The CD8-bearing T cells are called "killer" cells because they destroy the pathogen-infected cells.

discouraging false leads. Consequently, by the late 1970s, most investigators had abandoned the hunt, convinced that human retroviruses did not exist and further effort was fruitless. Although believed by some to be misguided, Gallo was not deterred. Because there were so many examples of retroviral-induced leukemia in animals, he was persuaded that an oncogenic human retrovirus might yet be discovered. This conviction was supported when, in 1978, Gallo and his collaborators found a retrovirus from gibbon apes to be the cause of the animal's lymphocytic leukemia. Subsequent studies showed the gibbon ape leukemia viruses could be transmitted to other primate species. If such transmission could occur among nonhuman primates, why should humans be excluded?

Two factors were key to identifying the first human retrovirus. First was the availability of a simple but very sensitive assay for retrovirus particles via detection of reverse transcriptase activity. Second, as noted above, was an ability to maintain human lymphocytes in cultures by addition of IL-2. The first hint of success came from analysis of cells from a patient with a T-cell lymphoma. Collaborators first established a cell line from this lymphoma by culturing tumor cells in the presence of IL-2. Subsequent analysis by Gallo's associate Bernard Poiesz showed measurable levels of reverse transcriptase, an indication that the cells were producing retrovirus particles. This interpretation was rapidly confirmed by electron microscopy, which revealed virus particles in both the cells and the surrounding medium. At the time, a standard approach for isolating viruses was to mix uninfected lymphocytes, which were easily maintained in culture, with tissue suspected of containing a virus, in anticipation that the virus would infect and produce progeny particles in the lymphocytes. Using this method, Gallo and associates were able to confirm that a retrovirus was transmitted from the T-cell lymphoma line to uninfected lymphocytes. In 1980, they reported isolation of the first oncogenic human retrovirus, which they called human T-cell lymphotropic virus type 1 (HTLV-I).

As work was progressing in the Gallo lab, researchers on the opposite side of the world were rapidly heading toward identification of their own human retrovirus. Their research was inspired by a 1977 report from Kiyoshi Takatsuki and his colleagues at Kyoto University of an unusual clustering of a fatal T-cell leukemia in adult patients.[7] Although living in a variety of locations in Japan, all of the patients had been born and raised on or near the southern island of Kyushu. Several features of this leukemia were distinct from known malignancies, prompting the investigators to consider it an entirely new disease, which they called adult T-cell leukemia (ATL). In their report, these investigators noted that the shared geographic origin of the patients with ATL suggested that either genetic background or infection by an oncogenic virus in that area could play a role in the disease. In the fall of 1981, just months after the published discovery of HTLV-I, Yori Hinuma and his Japanese colleagues reported ATL was probably caused by a retrovirus. By the end of that year, these researchers had isolated the infectious agent, which they called adult T-cell leukemia virus (ATLV). It was also noted that 25 percent of healthy adults in the Kyushu area had antibodies to the virus, indicating its wide prevalence in this region.

Although initially thought to be two different viruses, comparison of the genome sequences of ATLV and HTLV-I revealed that the viruses were actually one and the same, differing from each other only as expected of isolates from distinct regions of the globe. In a joint publication of 1982,[8] the American and Japanese investigators agreed to reserve the name HTLV-I for the virus, in recognition of the priority of its discovery, and named the disease ATL, as the Japanese had contributed most to establishing its unique identity. In that same year, a second human retrovirus was isolated in the Gallo lab, from cells of a patient with a benign T-cell leukemia. Called HTLV-II, this virus has a similar genome structure and 70 percent nucleotide sequence homology with HTLV-I. However, unlike HTLV-I, HTLV-II has not yet been associated convincingly with human pathology.

HTLV-I Is a Novel Oncogenic Retrovirus

Although the exact number is unknown, it has been estimated that 5 to 10 million people worldwide are infected with HTLV-I. About 5 percent of those infected develop one of two diseases—ATL, the fatal disease described above, or a neurodegenerative inflammatory disease called HTLV-I–associated myelopathy / tropical spastic paraparesis (HAM / TSP), typically associated with infections that occur late in life. Both diseases are characterized by long periods of viral persistence. In ATL, the virus is usually acquired in infancy by mother-to-child transmission during breastfeeding, but clinical manifestations do not appear until thirty-five to fifty years have passed. HAM is associated with infections that occur later in life via blood products, sexual transmission, or sharing of contaminated needles among drug addicts.

HTLV-I is a deltaretrovirus (see Table 2.1) belonging to a group that includes members infecting Old World monkeys, called simian T-lymphotropic viruses (STLVs). It is believed that HTLV-I arose from repeated cross-species transmission of common ancestors of STLV from these primates to humans 30,000 to 40,000 years ago, generating the various subtypes of the virus found in different parts of the world. HTLV-I is now endemic (prevalent) in Africa, Melanesia, Japan, the Caribbean, and Central and South America. We believe the virus was first acquired in Africa and brought to the Asian continent and onward to the Americas by the migration of prehistoric people across the Bearing Strait to North America and the edge of South America. Strains originating in Africa may also have been brought to the Caribbean and South America with the slave trade between the sixteenth and nineteenth centuries. The origin of HTLV-I in Japan is uncertain. Recent data also support the idea that the virus was carried to Japan during human migration sometime before 300 BC, but contamination from early traders and their African servants is another possibility. Transmission to humans, through bites from nonhuman primates or slaughtering of such animals, still occurs in Africa, leading to the emergence of new HTLV-I infections.

While HTLV-I has the potential to infect a variety of human blood cells, its main target is the T lymphocyte characterized by a protein on its surface called CD4. The retroviral particles are not very stable and are, therefore, transmitted most efficiently by contact of an infected cell with one that is not infected. However, HTLV-I is replication competent; its genome includes fully functioning *gag-pol-env* genes. In addition, sequences downstream of *env,* in a region called "X," encode unique viral regulatory proteins. Unlike the transducing retroviruses, HTLV-I genomes do not carry captured cellular genes. Furthermore, while a provirus is found in the same location in all leukemic cells from a given patient, indicating that the cancer is derived from a single infected cell, the absence of preferred integration sites eliminates insertional mutagenesis as a likely mechanism of oncogenesis.

Some progress has been made in treating patients with ATL with antiviral drugs or by bone marrow transplant, but there is still no effective cure. The disease is usually fatal within a year of diagnosis. The extraordinarily long latency and relatively low incidence of leukemia among infected individuals has made it difficult to determine the exact mechanism by which HTLV-I causes this disease. Because neither

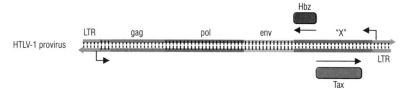

Fig. 5.5 HTLV-I provirus and genes encoding viral oncoproteins. A map of the HTLV-I provirus shows the position of sequences encoding the common retroviral genes *gag, pol,* and *env* (see Figure 2.4) and a unique region "X." The Tax and Hbz oncoproteins, as well as several others thought to contribute to the development of cancer, are encoded in X.

oncogene capture nor insertional mutagenesis is applicable, some viral gene or genes must be responsible for the development of this leukemia. Attention focused rapidly on the unique X region, found to encode several proteins with activities likely to contribute to the development of this disease. Of these, two proteins, Tax and Hbz, are central players and considered the key viral oncoproteins.

Tax is a multifunctional protein required for optimal expression from the HTLV-I proviral LTR. However, cell culture studies showed Tax also causes overexpression of a number of the cellular genes controlling T-cell physiology. The interaction of Tax with some cellular proteins promotes proliferation, and contact with others causes defects in DNA repair. This double whammy is thought to contribute to the genome instability and numerous chromosomal abnormalities characteristic of ATL. Given these important properties, it was somewhat surprising to find that the *tax* gene is absent or shut down in almost half of patients with ATL. The finding suggested that Tax protein might be required for initiation of the leukemia but not its maintenance, an idea confirmed with the discovery of the second critical protein, Hbz. In 2002, Jean-Michel Mesnard and his coworkers at the CNRS in Montpelier France showed that the *hbz* gene encoded a nuclear protein regulating the synthesis of mRNA from both retroviral and cellular genes. Later studies showed *hbz* mRNA and protein production in *all* ATL cells. Moreover, although antagonistic to some Tax functions, Hbz activates genes in some of the same signaling and proliferation pathways required for leukemic cell maintenance. These findings have encouraged recent efforts to produce a vaccine against Hbz to treat or possibly prevent ATL.

Although the one-two punch of Tax and Hbz is now considered critical for the initiation and maintenance of ATL, additional proteins encoded in alternatively spliced, X mRNAs are also likely to contribute to pathogenesis. While absence of some of these proteins has no obvious effect on virus propagation in cell culture, they are required for replication and persistence of the virus when injected into rabbits or nonhuman primates. The long latency of ATL, as well as inability to monitor early disease progression, has made it extremely difficult to evaluate the effects of all of these viral oncoproteins. These limitations and other challenges to research with HTLV-I continue to confront scientists and physicians in attempts to understand and deal effectively with this retroviral-induced disease.

Updates from the Battlefield

The advocacy of several influential individuals, including the American philanthropist Mary Lasker, together with early encouraging results from the NIH Virus Cancer Programs of 1964 and 1968, were important selling points for passage of the U.S. National Cancer Act of 1971, the beginning of President Nixon's "War on Cancer." Advances in genetics and molecular biology led many to think it would only require concerted effort to rapidly (1) apply knowledge gained from the oncogenic animal viruses to cure the human disease and (2) identify and deal with viruses causing human cancer. With generous funding from the NIH and a mad scramble of investigators ready to enlist in the battle, it appeared victory must be just around the corner. Almost half a century later, this battle continues, with many successful skirmishes, but not yet what can be called triumph.

We know now that human tumor cells generally contain numerous abnormal genetic changes. Today, the number of proliferation-inducing genes that show increased activity in cancer greatly exceeds the number captured or activated by the oncogenic retroviruses of chickens, mice, and other animals. In addition, study of the DNA tumor viruses and some retroviruses has identified other, equally relevant cellular genes called "tumor suppressors." If oncogenes can be compared to the gas pedals in an automobile, then the tumor suppressor genes are analogous to brakes. Malfunction in either case results in uncontrolled cellular proliferation, the hallmark of cancer. Despite these complexities, human counterparts of certain retroviral oncogenes are acknowledged to be important "drivers" of some cancers, which appear to be dependent on a single dominant genetic change.

For example, the nuclear regulatory Myc protein, one of the most frequently deregulated proteins in human cancer, is a key driver in certain B-cell lymphomas and human brain tumors. Her2, another driver, is an outer membrane protein closely related to that encoded in *v-erbB*, an oncogene first identified in a retrovirus of chickens that induced an acute form of leukemia called erythroblastosis. Increased expression or mutation of the human Her2 gene is a driver of some forms of lung cancer and an aggressive type of breast cancer. Indeed, antibodies targeting Her2 are used to treat patients with this cancer type. The driver oncogene *abl* was first identified in a mouse transducing retrovirus, the Abelson murine leukemia virus. A reciprocal exchange between human chromosomes 9 and 22, which causes the human *abl* sequences to be fused to those of another gene, *bcr*, is a hallmark of chronic myelogenous leukemia. The shortened chromosome 22 is known as the Philadelphia chromosome (in recognition of the city in which it was discovered). The translocation causes unbridled production of a fused Bcr-Abl protein, which, like Src, is an enzyme that activates other proteins by placing phosphates on specific tyrosine residues. The five drugs currently used to treat this disease are all inhibitors of the Bcr-Abl protein. These few examples illustrate some of the ways that research with animal retroviruses has enhanced our understanding of the genetic basis of cancer and influenced treatment of human disease. Analyzing the status of oncogenes and tumor suppressor genes has become a standard diagnostic tool for distinguishing among certain human malignancies and, in some cases, to suggest appropriate therapies in what has come to be called the practice of "precision medicine."

The inability to identify a human oncogenic retrovirus soon after the War on Cancer was declared led some investigators to grumble that the conflict had been launched for the wrong reasons. Discovery of HTLV-I quelled such complaints. HTLV-I was not only the first human retrovirus to be identified, but it was also found to have truly unique properties. Although cures for most cancers remain elusive, many insights into the biology of human cells have been garnered from the study of HTLV-I. Furthermore, a role for viruses in human cancers is no longer in doubt. It is currently estimated that viral agents are responsible for approximately one-fifth of the global human cancer burden. HTLV-I is the only known cancer-causing retrovirus. However, almost 700,000 deaths from liver damage and cancer each year are attributed to infection with the hepatitis B virus. The genome of this small DNA virus is also replicated via reverse transcription. Although a vaccine for hepatitis B virus has been available since 1981, access has been limited in parts of the developing world where up to 10 percent of the population is chronically infected. The human papillomavirus, another small DNA virus first connected to skin cancer in the 1970s, was identified as the major cause of cervical cancer in 1983 and recently associated with other malignancies, including head and neck cancer. A vaccine for this virus (HPV vaccine) has been available for immunization of children since 2006, but to date it has been underutilized. A third, related human DNA virus is the cause of a very rare skin cancer called Merkel cell carcinoma, for which surgical removal of early lesions is the only effective treatment.

Named after its discoverers, Michael Anthony Epstein and Yvonne Barr, the human Epstein-Barr virus is a member of the herpesvirus family, all of which contain DNA genomes encoding more than 100 viral proteins. The virus is best known as the cause of infectious mononucleosis (the kissing disease), but most residents of the United States will have been infected with the Epstein-Barr virus at some point in their lifetime, with little consequence. In 1968, Epstein-Barr virus was reported to cause infected B cells to grow indefinitely, providing the first indication of its oncogenic potential. The virus has since been associated with several forms of cancer, including Hodgkin lymphoma, Burkitt lymphoma, and gastric and nasopharyngeal cancers. Approximately 150,000 deaths worldwide were recently attributed to such malignancies. Another human herpesvirus, Kaposi sarcoma virus, is associated with a cancer common in patients with AIDS. A third member of this family, the human cytomegalovirus, has been associated with salivary gland cancers. There are currently no effective vaccines or cures for these herpesvirus infections.

The most widespread and deadly cancer virus, the hepatitis C virus, was discovered in 1989. This virus contains a single-stranded RNA genome encoding six viral proteins. Hepatitis C is most commonly transmitted from infected individuals by transfusion of contaminated blood or by contaminated needles of drug users. Hepatitis C virus is estimated to infect up to 150 million people worldwide and accounts for approximately a million liver cancer deaths every year. There is no vaccine for hepatitis C. While curative drugs have become available in the past few years, they are prohibitively expensive in many places.

Although the battle is not yet over, the many accomplishments and limited successes of the NIH Virus Cancer Program and Nixon's War on Cancer are certainly worth celebrating. Efforts to discern how retroviruses induce cancers in various animal species led to knowledge of the genetic basis of these diseases. In the process, researchers also acquired a deep understanding of the biology of oncogenic viruses. Effective methods for detection and characterization of such viruses were available barely a decade after the war began. A conviction that humans, like other animals, should be susceptible to retroviral infection was the accelerant for discovery of the first oncogenic human retrovirus, dispelling the notion that human beings are immune to such agents. However, as will be clear from what follows, one of their most important legacies of the War on Cancer was establishment of both the conceptual and technological underpinnings for response to the catastrophic emersion of what was to be the most deadly human retrovirus, HIV.

HIV and the AIDS Pandemic

Thanks to the widespread use of antibiotics and vaccines following World War II, infectious diseases seemed largely a thing of the past. A generation of "Flower Children" had renounced war and conventional society, proclaiming love for all. By the late 1970s, a sexual revolution was in full swing. Consequently, physicians in city hospitals were thoroughly bewildered by a mysterious and horrific new disease. An increasing number of previously healthy gay men were being admitted with a variety of opportunistic infections, including a rare lung infection, *Pneumocystis* pneumonia, and a type of cancer called Kaposi sarcoma, which was normally seen only in elderly men and rarely in the Western world.[1] By the time patients arrived in their wards, most were desperately ill, with bodies wasted and covered by lesions. All attempts at treatment failed. The best that could be offered was palliative care and morphine to ameliorate suffering. It appeared that a horrible new agent was playing havoc on the gay community.

The Centers for Disease Control and Prevention (CDC), the leading national public health institute of the United States, is responsible for the development and application of measures to control and prevent infectious disease. In May 1981, a report from Michael Gottlieb, a medical doctor and internist in Los Angeles, and his associates, caught the attention of James (Jim) Curran, the new chief of the Sexually Transmitted Disease Control Division at the CDC in Atlanta. The Los Angeles physicians reported that "in the period October 1980–May 1981, 5 young men, all active homosexuals, were treated for biopsy-confirmed Pneumocystis carinii pneumonia at three different hospitals in Los Angeles California."[2] His comments noted that these patients had virtually no CD4+ T (helper) cells but an abnormally high number of CD8+ (cytotoxic) T cells in their blood. These and other irregularities were indicators of severe immune system dysfunction. All five men eventually died of their affliction.

The CDC had just completed coordination efforts for two epidemics in the United States, Legionnaires disease and toxic shock syndrome. An astute epidemiologist, Curran suspected that the mysterious disease affecting gay men might foreshadow the eruption of another serious public health emergency. The observations from Los Angeles were published in the *New England Journal of Medicine* later that year, along with two similar reports from physicians in New York City. Learning of additional cases in other major cities, Curran immediately organized a task force of eager "epidemic intelligence officers in training" to review clinical records in the eighteen largest cities in the United States and determine how widespread this emergency might be. By the end of 1981, a total of 270 cases of Kaposi sarcoma or opportunistic infections associated with immune dysfunction had been documented among gay men. Almost half had died. However, it had become clear that the affliction was not limited to the gay male community. In December 1981, the first unusual cases of *Pneumocystis* pneumonia in individuals who inject drugs were documented. Six months later, in June 1982, the CDC published a report of unexpected, fatal opportunistic infections in men with hemophilia (the bleeding disease), and by the end of that year, four cases were documented in infants. Convinced that unusual instances of Kaposi sarcoma and / or life-threatening opportunistic infections were manifestations of the same affliction, the CDC task force named the disease AIDS (for acquired immune deficiency syndrome) and defined a case as "a disease at least moderately predictive of a defect in cell-mediated immunity, occurring in a person with no known cause for diminished resistance to that disease."[3] By 1983, AIDS was also detected in female partners of infected males.

Curran's tenure as coordinator of the CDC's Task Force on AIDS was initially anticipated to be completed within three months. As it turned out, Curran devoted the next fifteen years to fighting this disease. Based on the estimated number of gay men in the U.S. population, he later calculated that by 1981, when the first cases were coming to light, approximately 250,000 individuals in the community were already infected with the virus later identified as the cause of AIDS. Profoundly moved by the dreadful and hopeless plight of its victims and the frightening epidemiological implications, Curran became a passionate advocate for research and for development of preventive measures. With these imperatives in mind, he met with leaders at the National Institutes of Health (NIH) in Bethesda, Maryland, to stress the importance of finding the cause of AIDS. Identifying the responsible agent was an essential first step if there was to be any hope of controlling the looming epidemic.

Although there was much initial speculation, the cases among men with hemophilia suggested that a virus might be the culprit. People with this blood disorder lack an essential protein called clotting factor and are regularly treated with a concentrated, factor-containing product prepared from the blood of thousands of donors, some of whom might have been infected with such a virus. At the time, clotting factor preparations were known to be responsible for the viral hepatitis infections common in hemophiliacs. Therefore, it seemed likely that a virus responsible for AIDS was also transmitted in this way. Based on these considerations, in 1983, the CDC Task Force issued the first interagency recommendations for prevention of AIDS: sexual contact should be avoided with persons known or suspected of having AIDS, and members of groups at risk of developing AIDS should refrain from donating blood.

Discovery of the AIDS Virus

Several key investigators at the NIH were motivated by Curran's pleas for help, including Bob Gallo, who was flush with the success of discovering the first human retrovirus, human T-cell lymphotropic virus type 1 (HTLV-I). Although at first it seemed to Gallo that AIDS was just another "obscure disease of a small number of people," his interest was sparked when he learned that the same class of T cells targeted by HTLV-I was affected in AIDS.[4] Furthermore, the disease was transmitted in a similar way—through blood exchanges, sexual contact, and from mother to child. It seemed likely to Gallo that AIDS was also caused by a retrovirus, and he was convinced that an AIDS virus would be related to HTLV-I. In the spring of 1982, Gallo's laboratory began efforts to identify the virus based on this assumption. Gallo also organized a small group of scientists in related fields to focus on the disease and to chart the best directions for further research. Later to be called the Cancer Center AIDS Task Force, this group included Anthony (Tony) Fauci, an immunologist who would eventually play a central role in efforts to fight the disease, and Samuel (Sam) Broder, an oncologist who was to identify the first effective drug for the treatment of AIDS.

By the early 1980s, AIDS was also erupting among gay men in Europe. Aware of Gallo's recent discovery of human HTLV-I, French clinicians also suspected that a strain of this virus might be the cause of AIDS. To test this idea, they enlisted researchers at the Pasteur Institute in Paris who were studying oncogenic animal retroviruses. By February 1983, Francoise Barré-Sinoussi, a junior associate in Luc Montagnier's laboratory, obtained the first hint that a retrovirus was present in these patients. Using reagents and methods developed in Gallo's lab for identification of HTLV-I, she detected low amounts of reverse transcriptase activity in tissue obtained from a French patient with lymphadenopathy (an abnormal enlargement of the lymph nodes), an early sign of AIDS. By April, Montagnier's group had obtained clear evidence of retrovirus propagation in human T cells and observed retroviral particles budding from these cells by electron microscopy. They also showed that the retrovirus infected and killed lymphocytes from healthy donors. In addition, patients with AIDS produced antibodies that reacted with the retrovirus, a clear indicator of infection. Barré-Sinoussi presented findings from the Montagnier lab at the annual RNA Tumor Virus Meeting at Cold Spring Harbor in 1983. As their AIDS-associated retrovirus did not react with antibodies against HTLV-I, the French team concluded it must be a new retroviral species unrelated to HTLV-I. In their May 1983 publication, they named their retrovirus lymphadenopathy virus (LAV), because the first patient from whom the virus was isolated exhibited this symptom.[5]

As research progressed in France, investigators in the Gallo laboratory also saw hints of retrovirus presence in the AIDS patient samples they were testing. Their approach to preparing a stock was to combine blood samples from several patients with AIDS, with the expectation that the most vigorous viral strain would emerge from the infected lymphocytes—all the while assuming that the virus would be related to HTLV-I. Complicating their studies, it was discovered later that one or more of their Haitian patients with AIDS were also infected with HTLV-I (which is common in that Caribbean country). Having both viruses in their samples was a source of confusion—sometimes isolates seemed to be related to HTLV-I (as Gallo expected), but sometimes not. During the course of this work, Gallo learned of the French isolates and acquired a sample from Montagnier's group of what was thought to be LAV_{Bru} (named for their first patient, surname Brugiere). Unknown to the French team, this sample had become contaminated with one of their later isolates, LAV_{Lai}, obtained from a patient with Kaposi sarcoma. The contaminating virus grew very robustly and had therefore overtaken LAV_{Bru} during propagation. Indeed, LAV_{Lai} grew so well that it flourished in Gallo's isolation culture, which it also contaminated. Unaware of the mix-up and finding that the properties of LAV_{Lai} were distinct from those reported for LAV_{Bru}, Gallo assumed that this virus had originated from one of his patient samples. Thinking it to be unique but probably related to HTLV-I, he reported discovery of this "new" retrovirus in May 1984, calling it HTLV-III.[6]

The fact that contamination with LAV_{LAI} occurred in laboratories on both sides of the Atlantic may seem somewhat surprising. However, biological containment facilities were quite limited at the time. In most laboratories, the same incubators and working areas were used for making new viral isolates and for maintaining viral stocks through cycles of propagation. As virus propagation tends to select for strains that reproduce rapidly, such contamination was almost inevitable under those conditions. In addition, the pressure to identify the AIDS virus was intense, and competition to be first to succeed was keen. Unfortunately, it was not until some years later, when the entire nucleotide sequences of these viral genomes were compared, that this sad mix-up became apparent.

Three months after the report of HTLV-III, a third independent retrovirus isolation from patients with AIDS was reported from the laboratory of Jay Levy at the University of California in San Francisco. Aware of the LAV isolated by the French group, Levy began looking for a retrovirus in clinical specimens obtained from patients with AIDS in the San Francisco Bay Area. A former associate and collaborator, Paul Volberding, had been treating gay male patients with Kaposi sarcoma and was among the first to recognize that affliction with this rare cancer was a consequence of the fact that these patients' immune systems had been compromised. Subsequent testing of immune cells cultured from the blood of patients with AIDS provided the first positive results in the Levy lab and eventual isolation of an AIDS-related retrovirus, called ARV, in their publication of November 1984.[7]

One of the first indications that a retrovirus was not just "associated with" but actually *caused* AIDS was the demonstration by Levy's group that the agent could survive conditions used in preparation of the clotting factor vital to hemophiliacs. Their subsequent discovery that the retrovirus was inactivated by heat treatment, while still preserving clotting factor activity, saved the lives of many recipients at the time. The tragic infection of two U.S. laboratory workers during large-scale preparation of the AIDS retrovirus provided additional evidence that it was the cause of the disease. These incidents also emphasized the risks that exist for those who dedicate their lives to dealing with dangerous infectious agents. Finally, detection of the retrovirus in numerous individuals with AIDS by Gallo and others was sufficient to convince most in the scientific and medical communities that a retrovirus is, indeed, the cause of this disease.

Disputes and Contentions

What's in a Name? As noted above, by 1991, analysis of their genomes made it clear that LAV, HTLV-III, ARV, and other AIDS patient isolates were all closely related retroviruses, belonging to the same species. Although sense of ownership is strong, and the rights and input from discoverers needed to be respected, the plethora of names for this important pathogen began to generate unnecessary confusion in the biomedical field. A subcommittee empowered by the International Committee on the Taxonomy of Viruses was charged with choosing an acceptable name that could be used by all. At that stage of the pandemic, AIDS was a dreaded disease. Its unfortunate victims were stigmatized and shunned by many. In its deliberations, the committee asked a number of clinicians how a name that included the term AIDS might affect their patients. Although many respondents believed it would matter little, several felt that it would be easier to explain the difference between infection and the disease if the name were less pejorative. Accepting this reasoning, the committee avoided inclusion of the term AIDS and chose a name based on the agent's principal physiological effect: human immunodeficiency virus (HIV). In retrospect, this was a propitious decision; thanks to currently available treatments, individuals infected by HIV may now be spared the ravages of AIDS.

Patents and Politics Discovery of HIV isolates in both France and the United States led to furious races to develop blood-screening tests for the presence of the virus. Such tests were urgently needed, not only to identify infected individuals but also to ensure the safety of public blood supplies, which comprise pools from many donors, any one of whom might be infected unknowingly. In the absence of such a screen, any person with a medical condition serious enough to require a blood transfusion was at risk of becoming infected with HIV. Sadly, before a blood test for HIV was available, a number of unfortunate individuals were to fall victim to the disease from transfusions with contaminated blood.

In the interest of public health (and anticipation of licensing fees projected to be worth millions of dollars), the Pasteur and NIH groups were both quick to file separate patents describing blood tests for their viruses. An application from the Pasteur Institute was filed in the United States in December 1983, and an application from the NIH was filed the following April, on behalf of the U.S. government. Despite its later submission, the NIH patent was awarded first, in May 1984. What followed was a scientific controversy in which patenting agencies, lawyers, businesses, and even national governments quickly became embroiled.

"FRENCH SUE U.S. OVER AIDS VIRUS DISCOVERY," screamed the *New York Times* on December 14, 1985. At issue was not only recognition that the French researchers were the first to discover HIV but also access to related commercial rights. The legal dispute was eventually settled in an out-of-court agreement two years later. Announced jointly by French prime minister Jacques Chirac and U.S. president Ronald Reagan in 1987, the agreement stipulated that Montagnier and Gallo would be recognized as co-discoverers of HIV. In addition, the U.S. government agreed to allow a Pasteur Institute licensee to market a blood-screening test in the United States without legal challenge and to share royalties collected by the U.S. government for sales of such tests by its licensees. While these legal battles may seem petty compared to the number of people dying of AIDS, the controversy helped to bring worldwide attention to this virus and likely accelerated use of tests to ensure the safety of the blood supply.

The Gallo laboratory's ability to culture large quantities of interleukin-2 (IL-2)–supported T cells allowed mass production of HIV and rapid development of an accurate and sensitive blood-screening test. The test was based on the recognition that antibodies capable of reacting with retroviral proteins are only produced in response to viral propagation in an infected individual. Retroviral proteins separated by size and affixed to nitrocellulose strips were used to detect such antibodies in a patient's blood. The first assay of this type to be employed in clinical medicine, it was quickly adopted by physicians and in blood banks worldwide. Use of this assay allowed identification of infected individuals even before they exhibited symptoms of AIDS, alerting them and their sexual partners of the need to adopt preventive measures. The assay was also instrumental in safeguarding the blood supply, a major public health contribution.

Although legal matters were settled by 1987, the revelation in 1991 that Gallo's HTLV-III was actually the French LAV$_{LAI}$ led to highly publicized (and some quite acrimonious) recriminations. Some observers even questioned the scientific integrity of Gallo and his colleagues until it became clear that Montagnier's own virus cultures had become contaminated with LAV$_{LAI}$. After a protracted and thorough official inves-

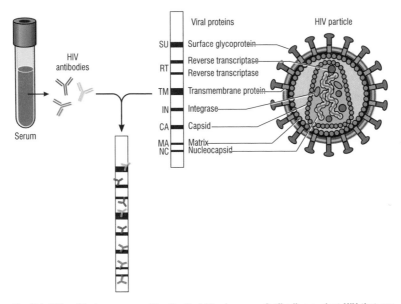

Fig. 6.1 HIV antibody assay provides the first blood screen. Antibodies against HIV that are present in the blood of infected individuals are identified by their ability to bind the retroviral proteins separated in test strips of nitrocellulose.

tigation, Gallo and his associates were cleared of malfeasance. He and Montagnier eventually agreed publicly that exchanges and contributions from both sides were critical to their individual successes. They acknowledged that Montagnier's group was the first to isolate and identify the retrovirus to be known as HIV. At the same time, Gallo and his colleagues were recognized for developing the reagents and culture methods that made its isolation possible, for establishing that HIV is the cause of AIDS, for and developing a vital blood screen.

However, the residue of intercontinental rivalry remained, if not among the protagonists then for others on both sides of the Atlantic. In 2008, Françoise Barré-Sinoussi and Luc Montagnier were awarded the Nobel Prize in Physiology and Medicine "for the Discovery of HIV." The third recipient of this prize was not Gallo, as most might have expected, but rather Harald Zur Hausen, who was recognized for his "discovery that the human papilloma virus causes cervical cancer." While no one would question the merits of all three recipients, the omission of Gallo was a surprise to many. If the prize had been given only for HIV, Gallo certainly would have been the obvious third recipient, and the Nobel Committee might have brought a final end to the HIV controversy.

Naysayers and Conspiracy Theories By the end of the 1980s, almost everyone in the biomedical community had accepted that HIV is the cause of AIDS. However, a few individuals challenged that idea. Most prominent among the naysayers was UC Berkeley Professor Peter Duesberg, who had played a pivotal role in the identification of the *src* oncogene and was by then a distinguished member of the U.S. National Academy of Science. Duesberg had a reputation as a contrarian, having disputed the importance of oncogenes in cancer. In 1987, he published a paper arguing that HIV is essentially harmless and that the early drugs used to treat it were the cause of the disease. These and other of Duesberg's arguments were subsequently refuted in scientific journals, public meetings, the media, and by a mountain of compelling evidence showing that HIV does cause AIDS. Nevertheless, Duesberg persisted with his claims, stating in a 2011 publication "HIV is not a new killer virus."[8] As noted previously, scientific disagreements are not uncommon, but most are generally resolved benignly with time and accumulating evidence. Unfortunately, Duesberg's "AIDS denialism" is considered to have had truly tragic consequences. Because of his stature as a scientist and the intense public interest and fear of AIDS, his claims drew widespread attention. Sadly, some knowingly infected individuals took these claims as license to have unprotected sex, wantonly spreading the virus to partners. Denialism was also embraced by Thabo Mbeki, president of South Africa from 1999 to 2008. Duesberg was invited to serve on an advisory panel to Mbeki, who accepted the scientist's views and condemnation of antiviral drugs. Mbeki's ban of such AIDS drugs in public hospitals has been estimated to be responsible for the premature death of up to 365,000 people and for many preventable infections, including those of infants. Although the ban was eventually revoked and South Africa has the largest antiretroviral treatment program globally, the country still has more people infected with HIV than any other country in the world: approximately 12 percent of its population of 56 million.

A variety of bizarre conspiracy theories have been concocted about the origin of the AIDS pandemic, some by authors of sensational magazine articles and books. One theory, rumored to have originated with the Russian KGB, claimed that HIV was fabricated by genetic engineering in U.S. government laboratories at Fort Detrick. The virus was said to have been tested secretly on prisoners and to have spread, after their release, to the gay community and others. Another theory suggested that HIV was a U.S.-engineered virus intended for population control, including elimination of blacks, Hispanics, and homosexuals. Several of the conspiracies were built around vaccination efforts. One claimed that the smallpox vaccine somehow increased sensitivity to HIV. Another theorized that genetically engineered HIV incorporated in the hepatitis B vaccine from a U.S. vaccine trial was the cause of the outbreak. A widely circulated theory blamed the spread of AIDS in Africa to contamination of the oral polio vaccine. None of these theories were credible, and the polio vaccine story was soundly debunked by published evidence. Regrettably, such theories, even once dismissed by evidence, are far from harmless. They raise unwarranted barriers to acceptance of life-saving vaccines in the public sector. For example, in some areas of Nigeria, India, and Pakistan, promulgation of the fallacy that the oral polio vaccine causes infertility has thwarted the goal of worldwide eradication of the virus. In the developed world, a debunked claim of links to autism has persuaded some parents to reject vaccination for their children, causing a resurgence of potentially debilitating childhood diseases, such as measles and whooping cough.

Where Did the Pandemic Start?

Although the AIDS pandemic seemed to erupt suddenly and unexpectedly in the 1980s, it is now clear that the virus had been percolating in humans for almost a hundred years. As described in what follows, the story of its origin and wide-scale distribution is one that entails urbanization, social change, and the development of new transportation networks in colonial central Africa during the beginning of the twentieth century.

Shortly after the discovery of HIV, AIDS was found to be firmly established in the heterosexual population in Africa, where women were afflicted in equal proportion to men. In Africa, AIDS was known as the "slim disease" because its victims became ill, lost weight, and died, although no one knew why. Opportunistic infections and other manifestations of the disease were similar to those observed in the United States and Europe. On the other hand, homosexuality, drug use, and blood transfusions were not risk factors; rather, transmission via heterosexual intercourse accounted for most cases of AIDS in the African population.

Samples of HIV recovered from blocks of patients' tissue collected between 1959 and 1960 in Kinshasa (formerly Leopoldville), Democratic Republic of Congo, provided the first clues to when the virus emerged. Evolutionary biologist Michael Worobey and his colleagues compared the genome sequences of viruses from these samples with those of HIV isolated from the surrounding areas at later times. By calculating the rate of genetic change with time, they constructed a "molecular clock." Projecting backward with the clock, it was possible to extrapolate when the last common ancestor of all of these viruses might have existed. From these predictions and the pattern of spread to surrounding cities, they estimated that pandemic HIV evolved from an ancestor in Kinshasa sometime between 1900 and 1910.

Situated strategically near the Atlantic coast and on the Congo River, Kinshasa had become a colonial boomtown at that time. The nearly 3,000-mile-long river and its many tributaries flow through the world's largest tropical forest, connecting Kinshasa to eight countries and numerous cities. In the early 1900s, this liquid highway and the newly constructed roads and rail lines transported many thousands of people, including migrant male laborers, to major population centers and mining locations where sex work flourished. The virus, which had pre-

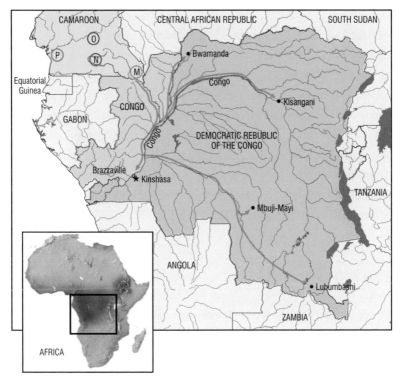

Fig. 6.2 Cradle of the AIDS pandemic. Central, sub-Saharan Africa is covered with one of the world's largest tropical forests, illustrated by the green portion in the map. As shown in the expanded section, the nearly 3,000-mile-long Congo River, with its tributaries, flows through this forest east and southward, connecting many neighboring nations. The circles at the top, left identify regions in Cameroon where the four distinct groups of HIV-1 emerged following transmission of SIVcpz and derivatives to humans from nonhuman primates. The region of origin of pandemic HIV-1 group M is indicated with a red circle, group O with blue, group N with purple, and group P with green.

viously spread slowly, virtually unnoticed, from one village to another for decades in the region around Kinshasa, now had ample chance to be spread to other regions. And spread it did. More than two-thirds of the world's nearly 40 million HIV-infected human beings live in sub-Saharan Africa.

The puzzle of how the virus came "out of Africa" to be dispersed throughout the world was also solved by Worobey and colleagues. Their examination of archival tissue samples from some of the earliest known Haitian patients with AIDS indicate that HIV arrived first in

Haiti. In the early 1960s, a large number of Haitian professionals were recruited to work in the newly independent Democratic Republic of Congo, where they were based in Kinshasa. One or more infected returning professionals could have initiated the subsequent AIDS outbreak in Haiti, the oldest epidemic center outside of sub-Saharan Africa. The next important stepping stone for HIV was New York City, an optimal location for subsequent worldwide distribution. Remarkably, genetic evidence suggests that this crucial move occurred via transmission of the virus by a single infected Haitian sometime around 1969. HIV had been circulating in the United States for about a decade before AIDS was first detected in 1980.

Where Did HIV Come from?

The first clue to the evolutionary origin of HIV came in 1986, from the discovery by Montagnier's group of a related but distinct retrovirus that causes a milder form of AIDS in western Africa. This virus was eventually called HIV type 2 (HIV-2) to distinguish it from the pandemic species, thereafter known as type 1 (HIV-1). Analysis of their genomes and morphological features revealed that both human viruses belonged to the genus *Lentivirus* (see Table 2.1). Lentiviruses, also called "slow" viruses, were long known to infect hoofed animals (horses, cows, goats, and sheep), causing blood and neurological disorders that develop over a protracted period of time. Although HIV-2 was only distantly related to HIV-1, it was closely related to a lentivirus isolated from captive macaques, in which it caused a fatal immunodeficiency.

Additional lentiviruses were subsequently discovered in numerous primate species in sub-Saharan Africa. Collectively known as simian immunodeficiency viruses (SIVs), they do not cause AIDS-like diseases in their natural hosts. This finding implied that the viruses have been associated with their respective host species for sufficient time to allow the relentless, virus-host genetic "arms race" to reach a mutually beneficial armistice. However, an SIV strain harmless in one primate can cause disease when transmitted to a different primate host species in which appropriate defense mechanisms have not evolved. For example, macaques are Asian monkeys and, unlike African monkeys, do not harbor SIVs. The HIV-2–related SIV strain isolated from the immunodeficient macaques described above was transmitted to these animals inadvertently. While in captivity, the diseased macaques had been inoculated with infected tissue samples from a natural SIV host, the African sooty mangabey monkey. Knowledge of this transmission suggested that human HIVs had been acquired in Africa by cross-species infection from resident primates. Based on sequence similarities, the source for HIV-2 is assumed to be the sooty mangabey monkey virus, known as SIVsmm (the suffix denoting species of origin). But where did pandemic HIV-1 come from?

Determining the origin of HIV-1 was a challenge that captivated Beatrice Hahn. During medical training in her native Germany, Hahn had been fascinated by the simplicity and elegance of the genetic code. Having worked with retroviruses for her thesis, she was delighted to join Bob Gallo's group for postdoctoral training. Arriving just after HIV-1 had been discovered, Hahn and George Shaw (another trainee and later also her spouse) were members of the team responsible for molecular characterization of the new virus. This experience laid the foundation for their continued investigations in independent but collaborating laboratories at the University of Alabama Medical School. Upon comparing the genetic sequences of a variety of HIV-1 isolates, Hahn became convinced that deducing the evolutionary origin of HIV-1 would require a deeper understanding of their relationships to the primate SIVs. The first informative sequence data came from analyses of SIVs isolated from animals housed in sanctuaries or primate centers. These data revealed that viruses isolated from chimpanzee (called SIVcpz) were most closely related to HIV-1. However, only three of the four sequences initially analyzed by Hahn and her colleagues seemed to fit into a clearly connected lineage with HIV-1. The sequence of SIVcpz from the fourth chimpanzee was puzzlingly different from the others.

A possible explanation for the diversity of sequences among SIVcpz isolates arose in a discussion with Hahn's British collaborator, Paul

Sharp, who had just learned that subspecies of chimps occupy geographically separate ranges in Africa. Sharp noted that the subspecies could be distinguished by analysis of their DNA. Exhilaration swept the Hahn group when their DNA tests showed that the three animals harboring SIVs most closely related to pandemic HIV-1 belonged to the very same subspecies of "central" chimpanzees, *Pan troglodytes troglodytes* (*P.t. troglodytes*), whereas the chimp with the most distant virus was from the "eastern" chimpanzee subspecies, *P.t. schweinfurthii*. Furthermore, the eastern chimps were believed to have diverged from the central subspecies several hundred thousand years ago, explaining why their SIVcpz should be so different. Because the natural range of the central chimpanzees coincided with the area in which HIV-1 is prevalent, Hahn and collaborators believed they had the answer. In 1999, they reported, with appropriate fanfare, that SIVcpz from *P.t. troglodytes* is the progenitor of HIV-1.

The idea that HIV-1 originated via cross-species transmission of SIV to humans from African chimps generated considerable excitement. Nevertheless, not everyone was convinced because the claim was based on analyses of viruses isolated solely from captive animals. Furthermore, extensive screening by numerous groups found that fewer than a dozen of the nearly 2,000 captive chimps were positive for SIV. It seemed possible that, as with the captive macaques, the SIVs isolated from the few infected chimps might have been acquired from some other primates during captivity. Hahn and her colleagues realized that their proposed origin of HIV-1 could only be tested by comparing the sequences of SIVs in free-living animals with the sequence of HIV in regions where the human virus was thought to have emerged. However, obtaining blood or tissue samples from wild chimpanzees was not an option. Not only would this be perilous (for investigator and animal), but it was also prohibited, as chimps are an endangered species. The inspiration for a possible solution came at a meeting with primatologists where Hahn learned that these researchers are able to screen for pathogens in primate feces and urine collected in the field.

With a possible solution in sight, Hahn's laboratory focused on finding conditions that would allow antibodies and nucleic acids in primate fecal samples to survive a journey from collection in the jungles of Africa to the United States. Their success in delineating such conditions and establishing accurate and sensitive diagnostic methods provided a unique, noninvasive way of collecting the required information. Hahn, together with Michael Worobey and the French retrovirologist, Martine Peeters (both of whom had prior experience in the field), went on to mount a collaborative, international effort to determine where and how cross-species transmission of SIV from wild chimps to humans had occurred. The work drew upon the expertise of virologists, evolutionary biologists, and conservationists, but its success depended vitally on the recruitment of help from local residents with skills in locating the feeding and nesting sites of wild chimps where urine and feces could be collected. Subsequent analysis of thousands of samples acquired at hundreds of sites throughout central and eastern Africa revealed that about 6 percent of the central chimpanzees and 13 percent of the eastern chimpanzees were infected with SIVcpz. Sequence comparisons showed no evidence of virus transmission to humans from the eastern chimps. In contrast, sequences of viruses from a distinct, geographically isolated community of central chimpanzees in south Cameroon clustered tightly with pandemic HIV-1 group M. These results provided convincing verification that transmission of SIVcpz from the central, *P.t. troglodytes* subspecies of chimpanzees was indeed the origin of pandemic HIV. Sometime in the early 1900s, the retrovirus began its initially slow journey to Kinshasa, then onward to central and southern Africa, and eventually the world. Central chimps in another isolated community in Cameroon were subsequently identified as the source of a second SIVcpz transmission, giving rise to the rare HIV-1 group N, estimated to have emerged later, in the 1950s.

Somewhat unexpectedly, analysis of thousands of fecal samples from other ape species revealed that SIVcpz had been transmitted from chimpanzees in Cameroon not only to humans but also to the western, lowland gorillas sharing the chimp's habitats. As with humans, SIVcpz is pathogenic in gorillas, causing an AIDS-like disease. Furthermore, a virus transmitted from infected gorillas to humans, called SIVgor, was identified as the source of two additional groups of HIV-1. The first,

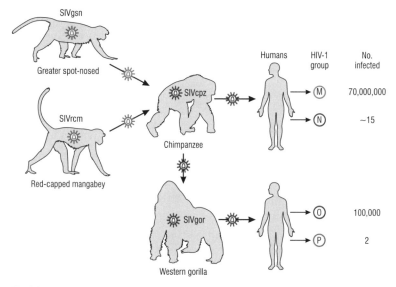

Fig. 6.3 HIV-1 origins via cross-species transmissions of SIV. The progenitor of pandemic HIV-1, the simian immunodeficiency virus of central chimpanzees (SIVcpz), is a product of recombination between two distinct SIVs that chimps contracted from monkeys on which they prey: red-capped mangabeys (SIVrcm, blue) and Cercopithecus monkeys such as the greater spot-nosed monkey (SIVgsn, red). SIVcpz (purple) was transmitted from chimps to humans and also to western gorillas in their habitats. HIV-1 groups M and N emerged from SIVcpz acquired from infected chimps, whereas HIV-1 groups O and P emerged from SIVgor, the SIVcpz-derived virus acquired from infected gorillas. The approximate number of infected humans is indicated at the right. Pandemic group M strains have infected many millions of people worldwide. Thousands have been infected with group O in Africa, but infections with groups N and P are rare.

HIV-1 group O, emerged around 1926 and has spread widely within Cameroon and the neighboring countries, infecting about 100,000 people. The date of emergence of the second, HIV-1 group P, is unknown. There are currently only two known cases of people infected with this retrovirus. The approximate geographic sites of origin for all four groups of HIV-1 are indicated in the map in Figure 6.2.

Subsequent studies revealed that both the central and eastern chimpanzees contracted their SIVs from other primates. Wild chimps hunt and eat monkeys, which are a natural reservoir for SIVs. Genomic analyses show that SIVcpz is a hybrid virus formed by recombination between SIV strains transmitted independently to chimpanzees from two species of monkey commonly on their menu. These SIVs are harmless in their natural monkey hosts. However, a long-term study of the wild chimp colonies in Gombe National Park, Tanzania (habituated to the presence of humans by Jane Goodall and her colleagues), revealed that SIVcpz is also pathogenic to chimpanzees. Infected animals have shorter life spans and immune cell deficiencies similar to those seen in humans with HIV-1. Although the exact manner in which SIVcpz entered the human population is unknown, the hunting of monkeys and apes for "bushmeat" is a likely explanation. Consumption of bushmeat not only is an accepted tradition but also provides important nutrition in Africa where protein sources are scarce. Nevertheless, capturing, butchering, and handling meat from these wild animals is apt to be a bloody affair, providing ample opportunity for SIVcpz, and indeed a plethora of other viruses, to enter a human host. As the harvesting of bushmeat continues to this day, such spillover of animal pathogens to humans, called "zoonoses," also continues.

A Most Formidable Adversary

More than three decades have elapsed since the discovery of HIV. Upward of 70 million people worldwide have been infected with the pandemic virus, and an estimated 35 million have died of AIDS. By almost any measure, the pandemic strain of HIV-1 (referred to simply as HIV hereafter) qualifies as one of the world's deadliest scourges. Remarkable achievements in medicine and public health measures have made it possible for many infected individuals to live nearly normal lives. Nevertheless, approximately 2 million people are newly infected each year, young adults predominant among them. Neither a protective vaccine nor a cure is yet available, and half of those infected lack accesses to antiviral drugs. While the virus can be transmitted from mother to child or by the contaminated needles of drug users, the vast majority (about 80 percent) of infections worldwide are acquired sexually. Owing to its intimate nature, controlling such transmission poses unique social and cultural complexities.

When acquired during sexual intercourse through semen, blood, or vaginal secretions, HIV passes through the outer layer of cells in the genital or anal mucosa, where it is picked up by local CD4+ T cells or other surveilling cells of the immune system. The virus is then transported to nearby lymph nodes in the body, sites where the normal adaptive immune responses to pathogens are initiated. Lymph nodes are packed with activated T cells, which display CD4 and other surface receptors to which the virus can attach, as well as cellular components needed for virus replication. Thousands of HIV particles are produced in a single infected T cell before it perishes in the process, within about a day. It is such robust viral propagation and cell killing that causes the lymphadenopathy and other flu-like symptoms characteristic of the first stages of infection. The virus then spreads to other sites in the body that harbor susceptible cells, including the spleen, thymus, and even the brain, where it can lead to dementia. A large proportion, 50 percent or more of the body's susceptible lymphocytes, are located in the intestine—positioned to defend against invasions by pathogens in the gut. Massive propagation of the virus and concomitant destruction of these gut-associated cells causes intestinal barriers to rupture, allowing microbes in the gut to enter the adjacent tissue. What ensues is a sustained stimulation of the immune system, which labors to address multiple calamities at the same time that a critical component of the system, the CD4+ lymphocytes, is destroyed.

During an initial "acute phase," at about three weeks after infection, virus production peaks while HIV-specific antibody production begins. However, the virus reproduces very rapidly and with genome-replicating enzymes that are error prone. Consequently, numerous antibody-evading viral mutants arise in each replication cycle. Because our adaptive immune system is not able to keep up with the ever-evolving virus, the disease enters a "persistent" or "chronic phase" in which a constant level of infectious virus particles (called the set point) is found in the blood. During both the acute and chronic phases, cells that do not support virus reproduction, such as long-lived quiescent T cells, are also infected. Such cells form a reservoir population that harbors "silent" proviruses in their genomes. Because no viral proteins are produced in these cells, they escape destruction by the immune system. However, the silent proviruses can be reactivated at later times, when quiescent cells are stimulated. The existence of these and other long-lived latent viral reservoirs, including infected cells in the brain and other organs, make the goal of eliminating HIV from the body an enormous challenge. Without antiviral treatment, the number of CD4+ T cells declines, often slowly but relentlessly. As these cells play a critical role in orchestrating the immune response, their decline imperils the patient's resistance to all infections. In the final stage of the disease, some five to ten years after infection, the CD4+ T-cell count is so low that even the most common, usually innocuous, infections can be lethal. At this point, HIV production again increases dramatically, and all of the devastation associated with AIDS is apparent.

HIV, like all viruses, depends on the proteins and other components in its host cell for propagation. Through eons of evolution and previous infections of common ancestors, primate cells (including those of humans) have acquired intrinsic defenses to block pathogen replication. Unfortunately, HIV possesses a number of countermeasures to defeat these cellular defenses. In addition to the common retroviral genes, HIV-1 (and its SIV progenitor) genomes contain six additional genes. Proteins encoded in two of these HIV-1 genes provide unique regulatory functions: one (Tat) enhances transcription of proviral DNA when cellular conditions are congenial, and the other (Rev) helps to ferry viral mRNAs from the nucleus to the cytoplasm for protein synthesis. The remaining four, called auxiliary genes, encode viral proteins that directly counteract cellular defense proteins. Two of the auxiliary viral proteins (Vif and Vpr) act early in the replication cycle to ensure viral DNA synthesis is not blocked. The other two (Vpu and Nef) promote efficient virus budding from the cell surface and remove surface proteins that mark infected cells for destruction. The auxiliary viral proteins all function as molecular adapters that bring defense proteins to the cellular machinery that normally destroys damaged or inactive proteins. In the natural hosts of SIVs, intrinsic cellular defense genes have evolved resistance to such retroviral countermeasures. As a result, the monkeys remain disease free, even though virus propagation persists—a state of equilibrium forged over millions of years. Because HIV is new to the human population, our intrinsic cellular defense

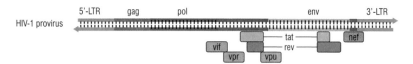

Fig. 6.4 Regulatory and auxiliary proteins encoded in the HIV-1 genome. The map shows the location of genes in HIV-1 proviral DNA. Names are acronyms for description of activities or effects initially observed for the proteins that they encode: Tat for transactivator for transcription, Rev for regulator of expression of viral proteins, Vif for viral infectivity factor, Vpr for viral protein affecting the rapidity of replication, Vpu for the fact that the viral protein is unique to HIV-1 and the related SIVcpz, and Nef for negative factor.

genes have not had sufficient time to adapt to the viral countermeasures. Furthermore, humans are not willing to wait for evolution to solve this problem.

Biomedical Science Fights Back

By 1984, when it first became clear that the AIDS-causing virus forms a lifelong association with its host, many clinicians and scientists were convinced that the disease would not be treatable with drugs. Among the few less doubtful was Sam Broder, then a medical oncologist at the NIH's National Cancer Institute. Broder was frustrated by an inability to offer any meaningful treatment to the desperately ill patients with AIDS who came to his and Tony Fauci's wards at the NIH Clinical Center. Many of the patients would die within just months of entry. The National Cancer Institute had established facilities for production and screening of compounds for anticancer activity. Broder was eager to apply this expertise to search for any drug that might be active against HIV. Even partial alleviation of the dreadful suffering of patients with AIDS seemed worth striving for. In September 1984, a meeting was held between Broder and virologists at the Burroughs Wellcome laboratory (subsequently GlaxoSmithKline) in Triangle Park, North Carolina, to discuss plans to test drugs as potential retroviral inhibitors.

Burroughs Wellcome was willing to join in the quest for anti-HIV drugs. The company had expertise in developing antivirals and just received Food and Drug Administration (FDA) approval for the drug acyclovir to treat herpesvirus infections. Acyclovir is a nucleoside analogue that includes one of the four normal bases in DNA (guanine, G). The herpesviral polymerase incorporates this drug in place of the normal substrate during DNA synthesis. Acyclovir is an effective antiviral because its incorporation blocks extension of the growing DNA chain, causing viral genome synthesis (and thereby viral propagation) to terminate. It seemed possible that a nucleoside analogue might also act as a chain terminator for DNA synthesis by retroviral reverse transcriptases.

HIV was considered too dangerous to bring into the Burroughs Wellcome laboratories. However, Marty St. Clair, a researcher in the company's virology group, had experience with other retroviruses. Thinking that an analogue inhibiting reverse transcription by a mouse retrovirus might also be effective against HIV, she set up a test using dishes containing single layers of cultured mouse cells. Each dish was treated with one of the company's collection of nucleoside analogues, after which a determined number of infectious mouse retroviral particles were added. In the absence of any drug, the mouse retrovirus killed the cells in which it replicated, producing holes (plagues) in the layer of cultured cells (illustrated in Figure 2.2). Promising compounds could then be identified by a decrease in the number of virus plagues. Success came late on a Friday afternoon in November 1984. St. Clair discovered that there were *no* plagues in nineteen of the more than 300 culture dishes she had set up that day. All nineteen had been treated with various concentrations of the same compound, a thymine (T)—containing nucleoside analogue called azidothymidine (AZT). It was a watershed moment. Word spread quickly within the company and to collaborators at the National Cancer Institute. It was now essential to determine if AZT would also be effective against HIV.

Broder and his NIH colleagues, including Bob Gallo, had established assays to test the effects of AZT and other analogues on HIV infection of human cells in culture and in the clinic. In a landmark paper of 1985, the Burroughs Wellcome / NIH collaborators reported that AZT inhibition of HIV reverse transcription blocks viral infectivity and cell killing at doses that do not severely impair cellular functions.[9] Subsequent clinical studies showed that drug treatment led to a reduction in the number of virus particles in the blood and measurable improvements in the immune function of patients with AIDS. There were signs of unacceptable toxicity with the drug, and some patients experienced serious side effects. But with no other treatment available, AZT was promptly approved by the FDA as the first antiviral for AIDS.

Sadly, the euphoria was short-lived. It soon became apparent that AZT-resistant mutants are among many produced in each cycle of HIV replication. Propagation of these mutants caused a resurgence of virus in the blood and relapse of the disease within relatively short periods of time. It was painfully clear that treatment with a single drug would not be sufficient to control the infection. Simultaneous treatment with two or more drugs would be required to reduce the probability of drug-resistant rebound. Nevertheless, the work with AZT was truly groundbreaking. The transient reduction in virus particles in the blood demonstrated to the nonbelievers that it is possible to block retrovirus reproduction by targeting viral proteins for inhibition. In addition, suppressing retrovirus production leads to reversal of some damage to the patient's immune system. The hint of success with one nucleotide analogue prompted an immediate search for others. Subsequent efforts in academia and pharmaceutical companies were spurred on by a vigorous and aggressive campaign by AIDS victims and their supporters (called ACT UP). In response to their demands for action and a growing appreciation of the problem's enormity, in the early 1990s, the NIH set aside 10 percent of its annual budget for the study of HIV and AIDS. And for the first time, nonscientist, patient advocates sat on the panels that reviewed AIDS research grant applications.

Broder and others pursued work with additional nucleoside analogues, but it would be four years until the next drug was approved for use in the clinic. More HIV reverse transcriptase inhibitors then became available, along with drugs that targeted the viral protease and integrase enzymes or blocked virus entry into host cells. By the end of 1995, the use of highly active antiretroviral therapy (called HAART), comprising combinations of antiviral drugs, resulted in a precipitous drop in the number of AIDS fatalities among young adults in the United States, for whom it had become the leading cause of death.

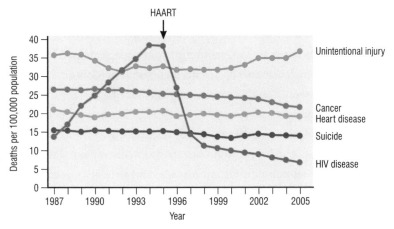

Fig. 6.5 Life-sparing effect of HIV antiviral treatment. Data reported by the U.S. Centers for Disease Control and Prevention in 2008 show that HIV infection had become the leading cause of death among young adults prior to introduction of highly active antiretroviral therapy (HAART), which comprised combinations of several antiviral drugs. A precipitous drop in such fatalities followed.

As of 2017, more than thirty anti-HIV drugs have been approved by the FDA, a larger number than available for any other virus. Current treatment comprises a combination of three or more drugs in a single pill to be taken daily. New formulations in the pipeline will need to be taken less frequently. Thanks to these medications, HIV infection no longer needs to be a death sentence. When diagnosed and treated early and consistently, the retrovirus can be reduced to undetectable levels in the blood, and patients may enjoy healthy lives of almost normal longevity. But while the infection can be held at bay with continuous antiviral treatment, HIV is not defeated. Failure to take the drugs will result in reemersion of the virus from the numerous reservoirs where proviruses lie waiting. Destruction of these reservoirs is therefore a much sought-after goal.

Among the millions of HIV-infected people in the world, only a single individual is known to have been cured of HIV infection. Timothy Ray Brown, an American living in Berlin, Germany, had been treated successfully with antiviral drugs for more than a decade when, in 2007, he was diagnosed with acute myeloid leukemia. Heroic procedures were undertaken to treat his cancer, including dosing with powerful cytotoxic drugs and radiation to kill all or most of his leukemic cells. This regimen was followed by transplantation of blood cells from a healthy donor to reconstitute his immune system. Hoping to "kill two birds with one stone," his Berlin physicians identified a donor with a genetic mutation that conferred resistance to certain strains of HIV. Their strategy was a success. Even though Brown stopped taking antiviral drugs following these procedures, there was no evidence of HIV in his blood, and his immune system gradually regained function. Unfortunately, all subsequent attempts to use similar procedures to treat other patients with AIDS have failed. Nevertheless, the sole example of Timothy Brown, known as "the Berlin Patient," has inspired hope that a simpler cure for HIV infection may someday be achieved.

Various practical measures to prevent HIV infection, such as the distribution of clean needles to drug users and promulgation of safe-sex practices, have been effective. Moreover, remarkable strides have been made in preventing infection from mother to child. Treatment with even a single agent such as AZT prior to delivery can prevent infection of newborns, and drug treatment during breastfeeding can eliminate HIV transmission through a mother's milk. Preexposure prophylaxis (PrEP), the use of antivirals by individuals who are not infected but at risk of contracting HIV (e.g., from an infected spouse), can also prevent infection. However, the most powerful and desirable weapon against HIV would be an effective vaccine.

In a hastily called 1984 press conference at the NIH, during which Bob Gallo reported his isolation of an AIDS virus, Margaret Heckler, then U.S. Secretary of Health and Human Services, predicted a vaccine would be available in two years. Caught up with the excitement at the time, it did not seem an unreasonable prediction, because an effective vaccine against a retrovirus causing leukemia in cats had just been developed. Little did anyone imagine that the world would still be waiting for an HIV vaccine more than three decades later.

Tony Fauci, director of the NIH's National Institute for Allergy and Infectious Disease since 1984, has been at the forefront in the fight against AIDS and in efforts to develop an HIV vaccine. As he has affirmed, this is still "one of the most formidable challenges facing scientists today."[10] Not only does the virus attack the very immune system that vaccines must engage, but its high mutation rate also produces strains that can evade antibodies capable of blocking viral infectivity. Furthermore, sugar molecules attached to the surface proteins of the virus cloak most sites that might otherwise be susceptible to such antibodies. While broadly acting antibodies that can block infection with many different strains of HIV were discovered in a very small number of people called "long-term survivors," these antibodies only appear years after the initial infection, when the virus has already established itself in the survivor's body. Nevertheless, some of these rare antibodies have shown potential as treatments or for possible long-term protection of individuals at risk of contracting HIV. Analysis of broadly acting antibody structures is providing critical insight into the features required for an effective vaccine or vaccination program. There is renewed hope that an HIV vaccine might be available in the foreseeable future.

Described as an "American hero" by President George W. Bush, Tony Fauci has also played a prominent role in the creation of the U.S. President's Emergency Plan for AIDS Relief (called PEPFAR), signed into law in May 2003. With a guaranteed $15 billion to be spent over five years, PEPFAR was the largest commitment ever made by any nation to such an international initiative. Through continued support by succeeding administrations, and in collaboration with partners in more than sixty countries, PEPFAR provided antiretroviral treatment to almost 12 million people in 2016 alone. Such efforts have made it possible for millions of babies with HIV-infected mothers to be born and remain free of the virus. With this encouraging progress, the Joint United Nations Programme on HIV / AIDS (UNAIDS) has set the ambitious goal of controlling the HIV pandemic by 2030, even in the absence of a vaccine. The plan is based on recognition that the pandemic can continue only if each infected person transmits the virus to at least one other person. If virus replication can be suppressed in 73 percent of the infected population, theoretically there will be enough breaks in the transmission chain to end virus spread. To reach its goal, UNAIDS established a strategy, based on diagnosis and treatment, by which this level of suppression can be achieved in all affected countries. It is calculated that the AIDS pandemic can be stopped if, by the year 2020, 90 percent of all people living with HIV will know their HIV status, 90 percent of all people with diagnosed HIV infection will receive sustained antiretroviral therapy, and 90 percent of all people receiving antiretroviral therapy will have viral suppression.

Distilled Knowledge

HIV has exacted an enormous toll on human beings. Too many people are still dying of AIDS. Nevertheless, much progress has been made in preventing and treating this disease. In the process, new knowledge about the HIV reproduction cycle and cells' intrinsic defense mechanisms has suggested additional approaches for the battle against this and other infectious agents. We also have a new appreciation for the origin of zoonosis. For example, recent analyses of primate fecal samples by Beatrice Hahn and associates, patterned on methods established for HIV, have revealed that *Plasmodium falciparum,* one of the most prevalent and lethal of the malaria parasites infecting humans, was originally acquired by humans from the western gorilla in Africa. Indeed, all known human strains of this parasite may have originated from a single cross-species transmission event. Knowledge of the human adaptive immune system acquired through the study of its response to HIV has suggested new strategies for treating not only this infection but other diseases as well. The progress made by biomedical science in its fight against AIDS has been nothing short of astonishing, all made possible by fundamental advances in molecular biology and the invaluable foundation established by years of research with retroviruses that infect animals. Without such tools and knowledge, we might still be wondering what causes AIDS.

Credit for the successes in AIDS treatment and prevention is due not only to biomedical science but also to the work of local public health and social workers who labor in some of the most hard-hit communities. In Southeast Asia and sub-Saharan Africa, where the disease burden is highest, response to this pandemic has bolstered ability to address subsequent zoonotic transmissions, including Ebola, the severe acute respiratory syndrome SARS virus, and others. In a world where a burgeoning human population continues to encroach on the natural environment, it is not a question of if but rather when the next pathogen spillover will occur. The HIV pandemic has provided valuable lessons to enhance future health security in our evermore interconnected globe.

— *Epilogue* —

Naturalists have long held that everything in our world is connected. As with a ball of wool, one only needs to pull on a single strand to eventually unravel the whole. That notion is no better upheld than in the story of the retroviruses. Such unraveling has led us back in evolutionary time, close to the very origin of life when both genes and enzymes were made of RNA. The ability of retroviral enzymes to reverse transcribe RNA into the more stable DNA and to insert that DNA into a host cell genome has been a major driving force in the evolution of life in our biosphere. Indeed, our own genome is largely a patchwork of remnants from these and other mobile elements. Some of the retroviral sequences in our genomes are artifacts from infections of our ancient ancestors. Many appear to comprise nonfunctional genetic debris. But as we have learned, others have been domesticated for critical steps in human reproduction. Some repurposing, such as the extraordinary exploitation of retroviral proteins in the evolution of mammalian placentas, is now fairly well understood. We also know that domesticated mobile elements orchestrated the genetic shuffling responsible for the diversity of antibodies that protect us from infections and for the production of surface molecules mediating communication between the cells of our adaptive immune system. Hints of additional roles for retroelements in human embryonic development, brain function, and disease imply that more examples of their influence are yet to be discovered.

The studies of oncogenic animal retroviruses and elucidation of the various ways in which they cause cancer have been instrumental in devising new approaches to treating human malignancies. Many therapies are based on the knowledge that the development of retrovirus-induced tumors depends on their introduction of, or control over, critical cellular genes. Such genes, called oncogenes and tumor suppressors, play important roles in pathways that regulate three essential processes: cell division and differentiation, cell survival, and genome maintenance. Changes in certain of these genes, called "drivers," are known to confer growth advantages for particular tumors. In some, albeit still limited cases, drugs targeting the proteins encoded in such drivers can be effective in treating malignancies. Large-scale sequencing of human cancers from different tissues and organs (e.g., breast, lung, colon) has been undertaken to identify driver genes that confer a selective advantage to diverse cancer cells. These studies have shown that every individual tumor is distinct with respect to its genetic alterations, but the signaling pathways affected in different tumors may be similar. Both features can be exploited. The genetic mutations produce protein changes unique to tumor cells, which can be targeted using novel immunotherapies, some to be described below. Drugs directed at pathways, rather than specific drivers, can also comprise effective therapies. The hope is that in the future, treatment for most cancers (and other diseases) will be determined on the basis of precise genetic or molecular profiling of the patient and his or her diseased tissue. We are just at the dawn of "precision medicine," when it will be possible to predict which treatment and prevention strategies will work for a particular individual or group of individuals.

Subverting the Enemy

Humans, however, have not been content with merely applying information gained from the study of retroviruses. The agents themselves are being exploited. Using recombinant DNA methods, retroviruses have been converted into delivery vehicles for directed gene transfer into almost any desired cell or organism. To produce such vehicles, the retroviral DNA is first gutted; viral genes (e.g., *gag, pol,* and *env*) are removed, leaving intact only important regulatory regions, including signals needed for reverse transcription and integration into a host genome. "Foreign" genes, or any chosen DNA sequences, are then inserted into what remains of the retroviral genome, now called a gene transfer "vector." In essence, the game plan is equivalent to what occurs in nature during formation of a transducing oncogenic retrovirus, except that the experimenter chooses the sequences that will be transduced. Particles containing such engineered retroviral genomes are then used to infect new host cells or, in some cases, whole organisms. When the engineered genome is integrated into the DNA of newly infected cells, the foreign DNA sequences may be expressed. Although no progeny particles can be formed from the gutted provirus, the infected cell (and all of its daughter cells) will have acquired new properties conferred by the inserted sequences.

Retroviral vectors are used routinely in research to study gene function in isolated cells or in cells within an experimental organism. Introduction of a foreign gene into an early embryonic cell can lead to the birth of what is called a "transgenic" animal, one in which every cell contains the added genetic information. For example, by inserting a jellyfish fluorescence gene into embryos using a retroviral vector, glow-in-the-dark, transgenic mice (and numerous other animals) have been propagated. Such transgenic animals allow researchers to observe the effects of various drug treatments or genetic manipulations within a living creature. In other applications, genetic information has been programmed for expression and analysis in particular tissues, such as in brain, liver, or other cells. The importance of a gene or genes to a normal (or diseased) organ can then be studied in a living animal.

The first agents to be exploited as gene transfer vectors were gammaretroviruses of mice. Many experiments with mice and other laboratory animals have since been performed with murine retroviral vectors. A French research team was the first to use such vectors in humans twenty years ago in a clinical trial to correct a genetic defect in children caused by a single gene mutation. As a result, the children were unable to produce T cells and other lymphocytes vital to fighting infections. They suffered from a condition called severe combined immunodeficiency (SCID). The only other treatments available to these vulnerable "bubble babies" produced harsh and often fatal side effects. For the gene therapy trial, lymphocyte precursor cells obtained from the children's bone marrow were infected with murine retroviral vectors that encoded a normal version of their mutated gene, after which the infected cells were infused back into the children's bodies. At first, the trial appeared to be a complete success; the children's immune systems were restored without any side effects. They no longer had to live in germ-free environments. But approximately two years later, one of the ten treated children developed leukemia. Analysis of the child's leukemic cells revealed that the retroviral vector caused the malignancy. In one of the patient's T cells, vector DNA had been integrated into the regulatory region of a gene that promoted cell proliferation. The insertion promoted uncontrolled expression of the gene and uncontrolled proliferation of this cell, leading to leukemia. Eventually three more cases of vector-induced leukemia via insertional mutagenesis arose among children in the French trial and one in a similar trial conducted in England. Sadly, one of the children died, but the four others recovered following treatment of their leukemias. While gene therapy trials were put on hold for a period of time, lessons from this experience led to the design of safer vectors and adoption of methods to reduce the probability of such insertional mutagenesis.

Human gene therapy trials are now ongoing in the United States and elsewhere. As incredible as it may seem, the vehicle of choice is not the murine retroviral vector initially engineered for this purpose but rather a lentivirus vector derived from the lethal pathogen, pandemic HIV. HIV-derived lentiviral vectors, gutted of all common and auxiliary viral genes, have become the preferred vehicles for gene therapy and a wide range of research applications. They have many advantages. Unlike the murine vectors, lentiviruses do not require a host cell to divide for integration of their DNA. Therefore, DNA of lentiviral vectors can be inserted into genomes of differentiated cells of the body (e.g., in muscle and brain), many of which divide infrequently, if at all. Furthermore, gene regulatory regions are not preferred sites for lentiviral DNA integration, reducing the potential for insertional mutagenesis. Consequently, lentiviral vectors experimentally supplemented with suitable surface proteins can be engineered to deliver genes with relative safety to a broad range of cell types, whether quiescent or actively dividing.

One of the first clinical applications of lentivirus vectors was reported in 2014. CD4+ T cells obtained from HIV-infected patients were infected with a lentiviral vector encoding genetic information that rendered the cells resistant to HIV. After placement back into the patients, immune functions were improved, HIV concentrations were reduced, and the HIV-resistant engineered T cells persisted in these patients, some for as long as five years. These and other positive results paved the way for additional clinical applications, including genetic manipulation of immune system cells to treat human cancers. In one method, T cells obtained from a patient with cancer are infected with a vector encoding an antibody-like surface protein (called a chimeric antigen receptor, abbreviated as CAR). The CAR protein is designed to bind a component unique to the patient's malignant cells, thereby initiating an immunological attack that destroys the cancer. Lentivirus-mediated CAR–T cell therapy has been approved for treatment of certain blood cell cancers and is being tested for other malignancies, as well as diseases such as hemophilia and sickle cell anemia. Genetic engineering of patients' cells to prevent or treat many different genetic and other diseases is a promising new frontier in precision medicine, with tools and approaches being created with breathtaking rapidity.

Directing Evolution

Recent, and currently the most accurate and versatile, tools for genetic manipulation are derived from genes encoded in Archaea and many bacteria. These single-celled prokaryotic organisms have harnessed gene-shuffling systems to acquire a unique form of immunity against their invaders. These systems go by the snappy name of CRISPR (which stands for the unwieldy "clustered regularly interspersed short palindromic repeats in DNA") and adjacent CRISPR-associated protein encoding genes, or simply CRISPR-Cas. The story of their discovery began in the 1980s, when several researchers noticed that the genomes of these single-cell microbes contain extensive arrays of short, repeated DNA sequences separated by seemingly random spacers of twenty-four to forty-eight base pairs. For almost twenty years, no one knew what to make of this peculiarity. When DNA sequencing technology became more readily available in the early 2000s, it was noted that such arrangements are widespread among these single-cell organisms, fueling speculation about functions they might encode. In 2005, researchers observed that the spacers between repeats in the arrays matched sequences in virus genomes infecting these single-cell microbes and suggested that the arrays could be components of a microbial defense system. Convincing evidence supporting this idea came from a somewhat unexpected direction.

Rodolphe Barrangou and Philippe Horvath, members of a research team at Danisco, a Danish food ingredient company, observed that some of the bacterial strains used to produce yogurt and other products were killed by certain bacteriophages, while others were resistant to these agents. Upon comparison of the bacterial strain sequences, they found that CRISPR arrays in the resistance bacteria genomes included spacers with sequences matching those in the genomes of bacteriophages to which they were immune. The arrays in sensitive bacteria lacked such spacers. Their experiments also showed that spacers with bacteriophage DNA sequences were acquired by the sensitive strains during the genesis of bacterial resistance, and that Cas proteins are required both for spacer acquisition and resistance to bacteriophage infection. The 2007 report of their findings in the journal *Science* launched a revolution in genetic experimentation that has had a profound influence on almost every field in biology and medicine.[1]

The mechanism by which spacers are acquired in CRISPR arrays in the microbial genomes has since been elucidated. When a new invader enters the cell, fragments of its genome are picked up by an assembly comprising two Cas proteins (Cas1 and Cas2) and then integrated into a repeat sequence at the beginning of the CRISPR array. This action produces a repeat-punctuated spacer "library" of sequences in the microbe's genome that can be used to identify and destroy the genomes of related future invaders. Some CRISPR-Cas systems deal only with invaders having DNA genomes. Spacer sequences are inserted via a mechanism similar to that of the retroviral integrase (illustrated in Figure 2.4), using the CRISPR repeat sequence as a target. In the process, the repeat is duplicated at either end of the acquired spacer, just as retroviral DNA integration creates a duplication of host target sequences at either end of a provirus. Other CRISPR-Cas systems can acquire immunity to invaders with RNA genomes, because their Cas1 protein includes a reverse transcriptase. The mechanism for spacer acquisition with these systems is similar to that deduced for human LINEs (illustrated in Figure 3.6). A segment of invader RNA is reverse-transcribed and the DNA product integrated into a terminal repeat, which is duplicated on either side of the acquired spacer DNA.

The next step in the process, defense against an invader, is accomplished by a mechanism that depends on one or more additional Cas proteins and an RNA molecule transcribed from the CRISPR array. One of the Cas proteins is a DNA cutting enzyme that destroys the invading genome but only when guided by a fragment of matching RNA derived from a spacer in the CRISPR library. The RNA fragment (called a guide RNA) binds to the Cas enzyme complex and also pairs with the invader's DNA in the region matching the RNA sequence. Dependence on the guide RNA for enzymatic attack accounts for the efficiency and specificity of these systems of immunity. The possibility that such microbial systems could be of value for practical applications, especially if engineered for use in other organisms, did not escape the attention of investigators. It was not long before the possibility became a reality.

Two teams of investigators showed that the CRISPR-Cas system from a human pathogenic bacterium, *Streptococcus pyogenes*, could be programmed to cut virtually any desired target DNA. Emmanuelle Charpentier and Jennifer Doudna at UC Berkeley reported in August 2012 that the system could be used to disrupt a targeted gene in another bacterium. They noted that CRISP-Cas systems offered potential for a variety of gene targeting and editing applications. A few months later, Feng Zhang and his associates at the Broad Institute and MIT showed that the system could be programmed to make precise cleavages at one or more sites in the genomes of both mouse and human cells. These researchers also demonstrated that a modified Cas DNA cutting enzyme could facilitate replacement of a targeted sequence within the genome of these cells. The reports from Charpentier, Doudna,[2] and Zhang[3] galvanized the biomedical world. Tools were now available to manipulate genomes, precisely and efficiently, in a vast array of organisms from microbes to plants, animals, and even humans.

CRISPR-Cas systems continue to be exploited and fine-tuned for genetic editing and control. Furthermore, incorporation of sequences encoding an editing enzyme and guide RNA into lentiviral or other vectors allows their introduction into almost any cell or organism. Biotech companies now construct single or multiple such lentiviral vectors. A researcher can specify the desired content and purchase appropriately engineered vectors for immediate use. Viral vectors encoding gene-editing systems have already been used in agriculture to create crop plants and farm animals with desirable properties. They have also been applied in laboratory research to produce organisms with a variety of genetic defects as models for human disease. In the clinic, the systems are being adopted for gene therapy. Some scientists have been experimenting with human embryos, where there is the potential to correct heritable diseases or affect other genetic changes at the earliest possible stage. Such potential has raised some thorny ethical questions as to what types of change may be acceptable.

As we have learned, much of the diversity and evolution of life on Earth has been created by transposable genetic elements with properties extant in present-day retroviruses. Such diversity has allowed selected genomes (winners) to survive and evolve through cataclysmic changes for more than three billion years. Although human civilization is only about 6,000 years old, our planet has now entered what has been called the Anthropocene epoch, in which human activities exert the dominant influence on our changing environment and climate. Many of the changes result from selective breeding of crops and animals needed to feed an exponentially expanding population, as well as the energy to fuel its activities. Associated encroachments on natural environments have accelerated the rate of extinction of numerous plant and animal species. It has been estimated that half of the animal species that once shared the Earth with us are already gone, with many others teetering on the edge of annihilation. Conversely, information gained in the past few decades has allowed humans to harness the genome-modifying capabilities applied by nature over eons of evolutionary history. We now have the capacity to thwart or prod evolution in numerous directions and in the proverbial blink of an eye. Gene editing is being tested as a way to free the environment from insect pests, such as disease-transmitting mosquitos. Use of gene transfer systems has been proposed as a method to restore needed diversity to dwindling, inbred animal populations and even for the resurrection of vanished creatures. Imagine the wooly mammoth back on Earth! Our new powers, informed by knowledge of the genome-shuffling capabilities exemplified by the retroviruses, offer previously inconceivable possibilities. There are many looming threats to our ecosystem to which such powers can be applied—hopefully in ways that do not harm but improve both human health and the world around us.

Final Reflections

Study of the retroviruses has spanned little more than a century. Their discovery coincided with the time when mysteries of heredity were being unraveled. Knowledge that the inheritance of form and function for each organism on Earth is encoded in DNA marked a watershed in science. The revelation that the code for translating this information is also conserved among all organisms in our biosphere confirmed that life on Earth has a common origin. Because physical traits are passed on faithfully from one generation to the next, it was thought initially that genomes must be very stable, with changes occurring in small, incremental steps. As elucidated by Darwin, organisms possessing advantageous properties were then selected by nature. The discovery that large segments of DNA can be moved within and between genomes by transposable elements, including retroviruses and their cousins, generated new appreciation of genetic dynamism and the significant role of retroviruses in evolution.

While it is still unclear if viruses arose before or after the first living cells, there is general consensus that the first hereditary material comprised self-replicating RNA molecules. However, viruses are the only currently existing biological entities with RNA genomes. The RNA viruses, along with functional RNA molecules within cells, are seen as remnants of this early "RNA world." Retroviruses provided unique insight into transition to the "DNA world" with discovery of reverse transcriptase, the enzyme that copies their RNA genome into a DNA molecule. The retroviruses' possession of a second enzyme, integrase, which inserts their DNA into the genome of the host cell, is unique among animal viruses. Integrated retroviral DNA, the provirus, is thereby ensconced as a permanent addition, programming the host cell machinery for production of progeny retroviral particles.

The ability of retroviruses to capture proviral-adjacent genes from their host's DNA first became apparent through research with cancer-causing animal retroviruses. Analysis of captured cellular oncogenes and related investigations of retrovirus biology disclosed the genetic basis of cancer. Oncogenes are now a major focus of cancer research, modern diagnosis, and treatment strategies. However, retroviral host gene capture is pertinent to evolution as well as cancer. Acquisition of an oncogene is readily apparent because it has easily observable consequences—namely, the rapid, unchecked proliferation of infected cells and induction of disease. But proviruses are integrated at numerous (essentially random) sites in the genomes of host cells. In most cases, they will be positioned close to genes for which capture is unlikely to produce such dramatic effects. Nevertheless, such proximal cellular genes are not expected to be exempt from retroviral capture. Collectively representing all sequences in the host genome, retroviral-mobilized cellular DNA can then be transferred to new locations in the host cell genome or to the genomes of other host species. Furthermore, upon infection of germ cells (eggs or sperm), these introduced sequences provide fodder for the evolution of new heritable functions in the recipient. It is now appreciated that exchange of genetic material, among organisms within and between all branches of the tree of life, has been facilitated throughout the eons by moveable genetic elements, including retroviruses and their progenitors.

The contributions of retroviruses and other retrotransposable elements to human evolution were brought home dramatically at the beginning of this century with the discovery that a major part of our own genome comprises DNA remnants of ancient retroviral infections and transpositions. While none of the human endogenous proviruses are capable of producing infectious particles, the physiological impact of some has been profound, affecting fetal development and other vital processes. Although not yet persuasive, hints of additional pathological effects are provocative, and we can expect more to be revealed as scientists continue to mine our genomes.

Retroviruses that infected various animal species were early appreciated as valuable models and tools for elucidating principles of genetics and the biology of cells on which their reproduction depends. However, for many years, they were thought to be irrelevant to human disease. That illusion was shattered with an understanding of the mechanisms by which oncogenic animal retroviruses caused cancers and the discovery of retroviruses, HTLV-I and HIV, that can cause fatal diseases in humans. The resources and intense focus of research on HIV in the fight against AIDS have greatly enhanced our understanding of retrovirus biology and its major target, the human immune system. Such knowledge has been exploited in the development of new strategies to treat HIV / AIDS and other diseases, some of which engage the very same retrovirus as a gene delivery vehicle.

Discoveries with retroviruses have led us back to the earliest appearance of life on Earth, through eons of evolution, and to the present day. Guided by illumination from these unique beacons, science has laid bare fundamental principles of heredity and evolution in our biosphere, and medicine is advancing in its battle with disease.

— *Notes* —

Chapter 1

1 P. Rous, "A Transmissible Avian Neoplasm (Sarcoma of the Common Fowl)," *J Exp Med* 12 (1910): 696–705, and "A Sarcoma of the Fowl Transmissible by an Agent Separable from the Tumor Cells," *J Exp Med* 13 (1911): 397–411.

2 A. E. Mirsky and A. W. Pollister, "Chromosin, a Desoyribose Nucleoprotein Complex of the Cell Nucleus," *J Gen Physiol* 30 (1946): 117–148.

3 A. D. Hershey and M. Chase, "Independent Function of Viral Protein and Nucleic Acid in Growth of Bacteriophage," *J Gen Physiol* 36 (1952): 39–56.

4 Francis Crick, *What Mad Pursuit* (New York: Basic Books, 1988), 76.

5 François, Jacob, *The Statue within* (Cold Spring Harbor, NY: Cold Spring Harbor Press, 1995), 264.

6 James W. Watson, *The Double Helix* (New York: Atheneum, 1968), 7.

Chapter 2

1 Franklin Stahl, ed., *We Can Sleep Later* (New York: Cold Spring Harbor Press, 2000), 9.

2 H. M. Temin and H. Rubin, "Characteristics of an Assay for Rous Sarcoma Virus and Rous Sarcoma Cells in Tissue Culture," *Virol* 6 (1958): 669–688.

3 D. Baltimore and R. M. Franklin, "The Effect of Mengovirus Infection on the Activity of the DNA-Dependent RNA Polymerase of L-Cells," *Proc Natl Acad Sci U.S.* 48 (1962): 1383–1390.

4 D. P. Grandgenett, A. C. Vora, and R. D. Schiff, "A 32,000-Dalton Nucleic Acid-Binding Protein from Avian Retrovirus Cores Possesses DNA Endonuclease Activity," *Virol* 89 (1978): 119–132.

Chapter 3

1 Alexander Rich, "On the Problems of Evolution and Biochemical Information Transfer," in *Horizons in Biochemistry*, ed. Michael Kasha and Bernard Pullman (New York: Academic Press, 1962), 114.

2 C. R. Woese and G. E. Fox, "Phylogenetic Structure of the Prokaryotic Domain: The Primary Kingdoms," *Proc Natl Acad Sci USA* 74 (1977): 5068–5090.

3 Sidney Altman, "Enzymatic Cleavage of RNA by RNA," Nobel Lecture, December 8, 1989.

4 Charles Darwin, *The Origin of Species* (New York: D. Appleton, 1859), 420.

Chapter 4

1 R. A. Weiss, "The Discovery of Endogenous Retroviruses," *Retrovirology* 3 (2006): 67.

2 E. Bosch, and M. A. Jobling, "Duplications of the *AZFa* Region of the Human Y Chromosome Are Mediated by Homologous Recombination between HERVs and Are Compatible with Male Fertility," *Hum Mole Genet* 12 (2003): 341–347.

3 P. J. Canfield, J. M. Sabine, and D. N. Love, "Virus Particles Associated with Leukemia in a Koala," *Aust Vet J* 65 (1988): 327–328.

4 Lewis Carroll, *Through the Looking Glass and What Alice Found There* (London: Macmillan, 1871), 41.

Chapter 5

1 Comments of Peter Vogt, http://library.cshl.edu/oralhistory/interview/cshl /memories/oncogenes-and-harry-rubins-laboratory/.

2 P. H. Duesberg and P. K. Vogt, "Differences between the Ribonucleic Acids of Transforming and Nontransforming Avian Tumor Viruses," *Proc Natl Acad Sci USA* 67 (1970): 1673–1680.

3 D. Stehelin, H. E. Varmus, J. M. Bishop, and P. K. Vogt, "DNA Related to the Transforming Gene(s) of Avian Sarcoma Viruses Is Present in Normal Avian DNA," *Nature* 260 (1976):170–173.

4 Leukemias are cancers of lymphocytes, immune cells localized in the bone marrow or in circulating blood. Carcinomas are cancers that originate in epithelial cells, such as found in the skin or the tissue that lines internal organs.

5 Lymphomas are tumors of immune cells that are localized to lymph nodes or other tissues, including the skin.

6 R. Dalla-Favera, M. Bregni, J. Erikson, D. Patterson, R. C. Gallo, and C. E. Croce, "Human c-myc onc Gene Is Located on the Region of Chromosome 8 That Is Translocated in Burkitt's Lymphoma Cells," *Proc Natl Acad Sci USA* 79 (1982): 7824–7927.

7 T. Uchiyama, J. Yodoi, K. Sagawa, K. Takatsuki, and H. Uchino, "Adult T-cell Leukemia: Clinical and Hematologic Features of 16 Cases," *Blood* 50 (1977): 481–492.

8 M. Popovic, M. S. Reitz Jr., M. G. Sarngadharan, M. Robert-Guroff, V. S. Kalyanaraman, Y. Nakao, I. Miyoshi, J. Minowada, M. Yoshida, Y. Ito, and R. C. Gallo, "The Virus of Japanese Adult T-cell Leukemia Is a Member of the Human T-cell Leukemia Virus Group," *Nature* 300 (1982): 63–66.

Chapter 6

1 An opportunistic infection is caused by a pathogen that does not normally harm the host but can cause disease in a host with weakened immunity.

2 M. S. Gottlieb, H. M. Schanker, P. T. Fan, A. Saxon, and J. D. Weisman, *Pneumocystis* Pneumonia—Los Angeles," *MMWR* 30 (1981): 1–3.

3 Centers for Disease Control (CDC), "Current Trends Update on Acquired Immune Deficiency Syndrome (AIDS)—United States," *MMWR* 31 (1982): 507–508.

4 Dr. Robert Gallo, interview by Dr. Victoria A. Harden and Dennis Rodrigues, *In Their Own Words—NIH*, 1994, https://history.nih.gov/nihinownwords/docs/gallo1_01.html.

5 F. Barre-Sinoussi, J. C. Chermann, F. Rey, M. T. Nugeyre, S. Chamaret, J. C. Gruest, C. Dauguet, C. Axler-Blin, F. Vézinet-Brun, C. Rouzioux, C. Rozenbaum, and L. Montagnier, "Isolation of a T-lymphotropic Retrovirus from a Patient at Risk for Acquired Immune Deficiency Syndrome (AIDS)," *Science* 220 (1983): 868–871.

6 R. C. Gallo, S. Z. Salahuddin, M. Popovic, G. M. Shearer, M. Kaplan, B. F. Haynes, T. J. Palker, R. Redfield, J. Oleske, B. Safai, G. White, P. Foster, and P. D. Markham, "Frequent Detection and Isolation of Cytopathic Retroviruses (HTLV-III) from Patients with AIDS and at Risk for AIDS," *Science* 224 (1984): 500–503.

7 J. A. Levy, A. D. Hoffmann, S. M. Kramer, J. A. Landis, J. M. Shimabukuro, and L. S. Oshiro, "Isolation of Lymphocytopathic Retrovirus from San Francisco Patients with AIDS," *Science* 225 (1984): 40–42.

8 P. H. Duesberg, A. Mandrioli, A. McCormack, J. M. Nicholson, D. Rasnick, C. Fiala, C. Koehnlein, H. H. Bauer, and M. Ruggiero, "AIDS since 1984: No Evidence for a New, Viral Epidemic—Not Even in Africa," *Ital J Anat Embryol* 116 (2011): 73–92.

9 H. Mitsuya, K. J. Weinhold, P. A. Furman, M. H. St. Clair, S. N. Lehrman, R. C. Gallo, D. Bolognesi, D. W. Barry, and S. Broder, "3′-Azido-3′-Deoxythymidine (BW A509U): An Antiviral Agent That Inhibits the Infectivity and Cytopathic Effect of Human T-lymphotropic Virus Type III / Lymphadenopathy-Associated Virus in Vitro," *Proc Natl Acad Sci USA* 82 (1985): 7096–7100.

10 A. Fauci and C. W. Dieffenbach, *NIH Statement on HIV Vaccine Awareness Day May 18, 2017,* https://www.nih.gov/news-events/news-releases/nih-statement-hiv-vaccine-awareness-day-2017.

Epilogue

1 R. Barrangou, C. Fremaux, H. Deveau, M. Richards, P. Boyaval, S. Moineau, D. A. Romero, and P. Horvath, "CRISPR Provides Acquired Resistance against Viruses in Prokaryotes," *Science* 315 (2007): 1709–1712.

2 M. Jinek, K. Chylinski, I. Fonfara, M. Hauer, J. Doudna, and E. Charpentier, "A Programmable Dual-RNA Guided DNA Endonuclease in Adaptive Bacterial Immunity," *Science* 337 (2012): 816–821.

3 L. Cong, F. A. Ran, D. Cox, S. Lin, V. Baretto, N. Habib, P. D. Hsu, X. Wu, W. Jianf, L. A. Marraffini, and F. Zhang, "Multiplex Genome Engineering Using CRISPR / Cas Systems," *Science* 339 (2013): 819–823.

— *Suggested Readings* —

Chapter 1

Books and Reviews

Crick, Francis. 1988. *What Mad Pursuit*. New York: Basic Books.

Maddox, B. 2002. *Rosalind Franklin, the Dark Lady of DNA*. New York: HarperCollins.

Watson, J. D. 1968. *The Double Helix*. New York: Atheneum.

Weiss, R. A., and P. K. Vogt. 2011. "100 Years of Rous Sarcoma Virus." *J Exp Med* 208 (12): 2351–2355.

Research Papers of Special Interest

Avery, O. T., C. M. Macleod, and M. McCarty. 1944. "Studies on the Chemical Nature of the Substance Inducing Transformation of Pneumococcal Types: Induction of Transformation by a Desoxyribonucleic Acid Fraction Isolated from Pneumococcus Type III." *J Exp Med* 79 (2): 137–158.

Creighton, H. B., and B. McClintock. 1931. "A Correlation of Cytological and Genetical Crossing-over in Zea Mays." *Proc Natl Acad Sci USA* 17 (8): 492–497.

Mendel, G. 1865. "Versuche über Plflanzen-hybriden" [Experiments concerning Plant Hybrids]. *Verhandlungen des naturforschenden Vereines in Brünn* [Proceedings of the Natural History Society] IV:3–47.

Meselson, M., and F. W. Stahl. 1958. "The Replication of DNA in *Escherichia Coli*." *Proc Natl Acad Sci USA* 44 (7): 671–682.

Chapter 2

Books and Reviews

Cairns, John, Gunther S. Stent, and James D. Watson, eds. 1992. *Phage and the Origins of Molecular Biology* (Expanded edition). Cold Spring Harbor, NY: Cold Spring Harbor Press.

Cooper, Geoffrey M., Rayla Greenberg Temin, and Bill Sugden, eds. 1995. *The DNA Provirus: Howard Temin's Scientific Legacy*. Washington, DC: ASM Press.

Crotty, Shane. 2001. *Ahead of the Curve: David Baltimore's Life in Science.* Berkeley: University of California Press.

Katz, R. A., and A. M. Skalka. 1994. "The Retroviral Enzymes." *Ann Rev Biochem* 63:133–173.

Skalka, A. M., and S. P. Goff, eds. 1993. *Reverse Transcriptase.* Cold Spring Harbor, NY: Cold Spring Harbor Press.

Svoboda, J. 2003. "Postulation of and Evidence for Provirus Existence in RSV-Transformed Cells and for an Oncogenic Activity Associated with Only Part of the RSV Genome." *Gene* 317:209–213.

Research Papers of Special Interest

Baltimore, D. 1970. "RNA-Dependent DNA Polymerase in Virions of RNA Tumor Viruses." *Nature* 226:1209–1211.

Beard, D., E. A. Eckert, T. Csaki, and D. G. Sharp. 1950. "Particulate Component of Plasma from Fowls with Avian Erythromyeloblastotic Leucosis." *Proc Soc Exper Biol Med* 76:533–536.

Bernhard, W., R. A. Bonar, D. Beard, and J. W. Beard. 1958. "Ultrastructure of the Viruses of Myeloblastosis and Erythroblastosis Isolated from the Plasma of Leukemic Chickens." *Proc Soc Exper Biol Med* 97:48–52.

Dulbecco, R. 1952. "Production of Plaques in Monolayer Tissue Cultures by Single Particles of an Animal Virus." *Proc Natl Acad Sci USA* 38:747–752.

Svoboda, J., P. Chýle, D. Šimkovič, and I. Hilgert. 1963. "Demonstration of the Absence of Infectious Rous Virus in Rat Tumour XC, Whose Structurally Intact Cells Produce Rous Sarcoma When Transferred to Chicks." *Folia Biol (Praha)* 9:77–81.

Temin, H. M., and S. Mizutani. 1970. "RNA-Dependent DNA Polymerase in Virions of Rous Sarcoma Virus." *Nature* 226:1211–1213.

Chapter 3

Books and Reviews

Blackburn, E. H. 1992. "Telomerases." *Annu Rev Biochem* 61:113–129.

Comfort, D. C. 2003. *The Tangled Field: Barbara McClintock's Search for Patterns of Genetic Control.* Cambridge, MA: Harvard University Press.

Craig, Nancy L., ed. 2015. *Mobile DNA III.* Washington, DC: ASM Press.

Hancks, D. C., and H. H. Kazazian Jr. 2016. "Roles for Retrotransposon Insertions in Human Disease." *Mobile DNA* 7:9. doi: 10.1186/s13100-016-0065-9.

Rich, A. 1962. "On the Problems of Evolution and Biochemical Information Transfer." In *Horizon in Biochemistry,* ed. M. Kasha and B. Pullman, 103. New York: Academic Press.

Woese, C. R. 1967. *The Genetic Code: The Molecular Basis for Genetic Expression.* New York: Harper & Row.

Research Papers of Special Interest

Belyi, V., A. J. Levine, and A. M. Skalka. 2010. "Unexpected Inheritance: Multiple Integrations of Ancient Bornavirus and Ebolavirus / Marburgevirus Sequences in Vertebrate Genomes." *PLoS Path* 6:e1001030.

Crick, F. H. 1968. "The Origin of the Genetic Code." *J Mol Biol* 38:367–379.

Guerrier-Takada, C., K. Gardiner, T. Marsh, N. Pace, and S. Altman. 1983. "The RNA Moiety of Ribonuclease P Is the Catalytic Subunit of the Enzyme." *Cell* 35:849–857.

Kazazan, H. H., Jr., C. Wong, H. Youssoufian, A. F. Scott, D. G. Phillips, and S. E. Antonarakis. 1988. "Haemophilia A Resulting from De Novo Insertion of L1 Sequences Represents a Novel Mechanism for Mutation in Man." *Nature* 332:164–166.

Kruger, K., P. J. Grabowski, A. J. Zaug, J. Sands, D. E. Gottschling, and T. R. Cech. 1982. "Self-Splicing RNA: Autoexcision and Autocyclization of the Ribosomal RNA Intervening Sequence of *Tetrahymena.*" *Cell* 31:147–157.

Miller, Stanley L., and Harold C. Urey. 1959. "Organic Compound Synthesis on the Primitive Earth." *Science* 130 (3370): 245–251.

Orgel, L. E. 1968. "Evolution of the Genetic Apparatus." *J Mol Biol* 38:381–393.

Chapter 4

Books and Reviews

Daugherty, M. D., and H. S. Malik. 2012. "Rules of Engagement: Molecular Insights from Host-Virus Arms Races." *Annu Rev Genet* 46:677–700.

Denner, J. 2016. Expression and Function of Endogenous Retroviruses in the Placenta." *APMIS* 124:31–43.

Katoh, I., and S. Kurata. 2013. "Association of Endogenous Retroviruses and Long Terminal Repeats with Human Disorders." *Front Oncol* 3: Article 234.

Ryan, F. 2009. *Virolution*. London: HarperCollins.

Ryan, F. P. 2016. "Viral Symbiosis and the Holobiontic Nature of the Human Genome." *APMIS* 124:11–19.

Weiss, R. A. 2006. "The Discovery of Endogenous Retroviruses." *Retrovirology* 3:67.

Xu, W., and M. V. Eiden. 2015. "Koala Retroviruses: Evolution and Disease Dynamics." *Annu Rev Virol* 2:119–134.

Research Papers of Special Interest

Blond, J.-L., D. LaVillette, V. Cheynet, O. Bouton, G. Oriol, S. Chapel-Fernandes, B. Mandrand, F. Mallet, and F.-L. Cosset. 2000. "An Envelope Glycoprotein of the Human Endogenous Retrovirus HERV-W Is Expressed in the Human Placenta and Fuses Cells Expressing the Type D Mammalian Retrovirus Receptor." *J Virol* 74:321–329.

Demogines, A., J. Abraham, H. Choe, M. Farzan, and S. L. Sawyer. 2013. "Dual Host-Virus Arms Races Shape an Essential Housekeeping Gene." *PLoS Pathog* 8 (5): e1002666.

Enard, D., L. Cai, C. Gwennap, and D. A. Petrov. 2016. "Viruses Are a Dominant Driver of Protein Adaptation in Mammals." *eLife* 5:e12469.

Hughes, J. F., and J. M. Coffin. 2001. "Evidence for Genomic Rearrangements Mediated by Human Endogenous Retroviruses during Primate Evolution." *Nat Genet* 29:487–489.

Levy, J. A. 1973. "Xenotropic Viruses: Murine Leukemia Viruses Associated with NIH Swiss, NZB, and Other Mouse Strains." *Science* 182:1151–1153.

Martin, M. A., T. Bryan, S. A. Rasheed, and A. S. Khan. 1981. "Identification and Cloning of Endogenous Retroviral, Sequences Present in Human DNA." *Proc Natl Acad Sci USA* 78:4892–4896.

Mi, S., X. Lee, X.-P. Li, G. M. Veldman, H. Finnerty, L. Racie, E. LaVallie, X. Y. Tang, P. Edouard, S. Howes, J. C. Keith Jr., and J. M. McCoy. 2000. "Syncytin Is a Captive Retroviral Envelope Protein Involved in Human Placental Morphogenesis." *Nature* 403:785–789.

Tarlinton, R. E., J. Meers, and P. R. Young. 2006. "Retroviral Invasion of the Koala Genome." *Nature* 442:79–81.

Chapter 5

Books and Reviews

Boveri, Theodor. 1914. *Zur Frage der Entstehung Maligner Tumoren.* Jena: Gustav Fischer.

Gallo, R. C. 2005. "History of the Discoveries of the First Human Retroviruses: HTLV-1 and HTLV-2." *Oncogene* 24:5926–5930.

Matsuoka, M., and K.-T. Jeang. 2007. "Human T-cell Leukemia Virus Type-1 (HTLV-1) Infectivity and Cellular Transformation." *Nat Rev Cancer* 7:270–280.

Rubin, H. 2011. "The Early History of Tumor Virology: Rous, RIF, and RAV." *Proc Natl Acad Sci USA* 108:14389–14396.

Varmus, H. 2017. "How Tumor Virology Evolved into Cancer Biology and Transformed Oncology." *Annu Rev Cancer Biol* 1:11.1–11.18.

Vogt, P. K. 2012. "Retroviral Oncogenes: A Historical Primer." *Nat Rev Cancer* 12:639–648.

Research Papers of Special Interest

Huebner, R. J., and G. J. Todaro. 1969. "Oncogenes of RNA Tumor Viruses as Determinants of Cancer." *Proc Natl Acad Sci USA* 64:1087–1094.

Neel, B. J., W. S. Hayward, H. L. Robinson, J. Fang, and S. M. Astrin. 1981. "Avian Leukosis Virus-Induces Tumors Have Common Proviral Integration Sites and Synthesize Discrete New RNAs: Oncogenesis by Promoter Insertion." *Cell* 23:323–334.

Payne, G. S., S. A. Courtneidge, L. B. Crittenden, A. M. Fadley, J. M. Bishop, and H. E. Varmus. 1981. "Analysis of Avian Leukosis Virus DNA and RNA in Bursal Tumors: Viral Gene Expression Is Not Required for Maintenance of the Tumor State." *Cell* 23:311–322.

Poiesz, B. J., F. W. Ruscetti, A. F. Gazdar, P. A. Bunn, J. D. Minna, and R. C. Gallo. 1980. "Detection and Isolation of Type C Retroviral Particles from Fresh and Cultured Lymphocytes of a Patient with Cutaneous T-cell Lymphoma." *Proc Natl Acad Sci USA* 77:7415–7419.

Vogt, P. K. 1971. "Spontaneous Segregation of Nontransforming Viruses from Cloned Sarcoma Viruses." *Virology* 46:939–946.

Chapter 6

Books and Reviews

Brown, T. R. 2015. "I Am the Berlin Patient: A Personal Reflection." *AIDS Res Hum Retroviruses* 31 (1): 2–3. https://doi.org/0.1089/aid.2014.0224.

Bushman, F. D., G. J. Naglel, and R. Swanstrom, eds. 2012. *HIV: From Biology to Prevention and Treatment.* Cold Spring Harbor, NY: Cold Spring Harbor Laboratory Press.

Centers for Disease Control. 1982. "Epidemiologic Aspects of the Current Outbreak of Kaposi's Sarcoma and Opportunistic Infections." *N Engl J Med* 306:248–252. http://dx.doi.org/10.1056/NEJM198201283060432.

Cohen, J. 1998. "The Duesberg Phenomenon." *Science* 266 (5191): 1642–1644.

Cold Spring Harbor Laboratory Archives. 2016. *HIV / AIDS Research: Its History & Future.* October 13–16, 2016. http://library.cshl.edu/Meetings/HIV_AIDS/.

Gallo, R. C., and L. Montagnier. 1987. "The Chronology of AIDS Research." *Nature* 326:435–436.

Research Papers of Special Interest

Gao, F., E. Bailes, D. L. Robertson, Y. Chen, C. M. Rodenberg, S. F. Nichael, L. B. Cummins, L. O. Arthur, M. Peeters, G. M. Shaw, P. M. Sharp, and B. H. Hahn. 1999. "Origin of HIV-1 in the Chimpanzee *Pan Troglodytes Troglodytes.*" *Nature* 397:436–444.

Gottlieb, M. S., R. Schroff, H. M. Schanker, J. D. Weisman, P. T. Fan, R. A. Wolf, and A. Saxon. 1981. "*Pneumocystis Carinii* Pneumonia and Mucosal Candidiasis in Previously Healthy Homosexual Men: Evidence of a New Acquired Cellular Immunodeficiency." *N Engl J Med* 305:1425–1431.

Mazur, H., M. A. Michelis, J. B. Greene, I. Onorato, R. A. Vande Stouwe, R. S. Holzman, G. Wormser, L. Brettman, M. Lange, and H. W. Murray. 1981. "An Outbreak of Community-Acquired *Pneumocystis Carinii* Pneumonia. Initial Manifestation of Cellular Immune Dysfunction." *N Engl J Med* 305:1431–1438.

Siegel, F. P., C. Lopez, G. S. Hammer, A. E. Brown, S. J. Kornfeld, J. Gold, J. Hassett, S. Z. Hirschman, C. Cunningham-Rundles, B. R. Adelsberg, D. M. Parham, M. Siegal, S. Cunningham-Rundles, and D. Armstrong. 1981. "Severe Acquired Immunodeficiency in Male Homosexuals, Manifested by Chronic Perianal Ulcerative Herpes Simplex Lesions." *N Engl J Med* 305:1439–1444.

Worobey, M., M. Gemmel, D. E. Teuwen, T. Haselkorn, K. Kunstman, M. Bunce, J. J. Muyembe, J. M. Kabongo, R. M. Kalengayi, E. Van Marck, M. T. Gilbert, S. M.

Wolinsky. 2008. "Direct Evidence of Extensive Diversity of HIV-1 in Kinshasa by 1960." *Nature* 455:661–664.

Worobey, M., M. L. Santiago, B. L. Keele, J.-B. N. Ndjango, J. B. Joy, B. L. Labamall, B. D. Dhed'a, A. Rambaut, P. M. Sharp, G. M. Shaw, and B. H. Hahn. 2004. "Origin of AIDS: Contaminated Polio Vaccine Refuted." *Nature* 428:820.

Worobey, M., T. D. Watts, R. A. McKay, M. A. Suchard, T. Granade, D. E. Teuwen, B. A. Koblin, W. Heneine, P. Lemey, and H. W. Jaffe. 2016. "1970s and 'Patient 0' HIV-1 Genomes Illuminate Early HIV / AIDS History in North America." *Nature* 539:98–101.

Epilogue

Reviews

Barrangou, Rodolphe, and Philippe Horvath. 2017. "A Decade of Discovery: CRISPR Functions and Applications." *Nat Microbiol* 2: Article 17092. http://dx.doi.org/10.1038/nmicrobiol.2017.92.

Cohen, J. 2017. "The Birth of CRISPR Inc.: How a Community Fractured as a Revolutionary Genome-Editing Tool became a Business." *Science* 355:680–684.

Collins, F. S., and H. Varmus. 2015. "A New Initiative on Precision Medicine." *N Eng J Med* 372:793–795.

Jackson, H. J., S. Rafiq, and R. J. Brentjiens. 2016. "Driving CAR-T-cells Forward." *Nat Rev Clin Oncol* 13:370–383.

Monastersky, R. 2015. "The Human Age." *Nature* 519:144–147.

Naldini, L., D. Trono, and I. M. Verma. 2016. "Lentiviral Vectors, Two Decades Later. A Deadly Virus became an Effective Gene Delivery Tool." *Science* 353:1101–1102.

Vogelstein, B., N. Papadopoulos, V. E. Velculescu, S. Shou, L. A. Diaz Jr., and K. W. Kinzler. 2013. "Cancer Genome Landscapes." *Science* 339:1546–1558.

Research Papers of Special Interest

Cavazzana-Calvo, M., S. Hacein-Bey, G. de Saint Basile, F. Gross, E. Yvon, P. Nusbaum, F. Selz, C. Hue, S. Certain, J.-L. Casanova, P. Bousso, F. Le Deist, and A. Fischer. 2000. "Gene Therapy of Human Severe Combined Immunodeficiency (SCID)-X1 Disease." *Science* 288:669–672.

— *Acknowledgments* —

I am most grateful to the Fox Chase Cancer Center in Philadelphia, my scientific home for thirty years, for the resources required to compose this volume. I also thank Marie Estes, my ever-willing and excellent administrative assistant, for helping to prepare it, and my husband, Rudy Skalka, and colleagues Glenn Rall and Jane Flint for reviewing the text and providing valuable suggestions and comments. Illustrations were produced with the skillful assistance of Patrick Lane, ScEYEnce Studios.

— Index —